For Gary

BEYOND THE BRAIN

BEYOND THE BRAIN

How Body and Environment Shape Animal and Human Minds

Louise Barrett

PRINCETON UNIVERSITY PRESS
PRINCETON AND OXFORD

Published by Princeton University Press, 41 William Street,
Princeton, New Jersey 08540
In the United Kingdom: Princeton University Press, 6 Oxford Street, Woodstock,
Oxfordshire OX20 1TW
press.princeton.edu

Second printing, and first paperback printing, 2015
Paperback ISBN 978-0-691-16556-1

The Library of Congress has cataloged the cloth edition of this book as follows

Barrett, Louise.
Beyond the brain : how body and environment shape animal and human minds /
Louise Barrett.
p. cm.
Includes bibliographical references and index.
ISBN 978-0-691-12644-9 (hardback)
1. Brain—Evolution. 2. Evolution (Biology) 3. Ecology. I. Title.
QL933.B27 2011
591.5—dc22

2010048477

British Library Cataloging-in-Publication Data is available

This book has been composed in Perpetua
Printed on acid-free paper. ∞
Printed in the United States of America

10 9 8 7 6 5 4 3 2

CONTENTS

ACKNOWLEDGMENTS

The nice thing about writing a book, if you're a fan of distributed cognition, is that it lets you practice what you preach. This book could never have been written if I hadn't borrowed all the brains that I could, to echo Woodrow Wilson. Some of the brains I borrowed were distributed in other books and articles, but there is also a more embodied means of exploiting other people's brains as a resource, and I have been very lucky to find myself in an environment full of smart, enthusiastic people who were more than happy to share their expertise and knowledge with me face-to-face.

First, thanks to my editors at Princeton, Alison Kalett, who saw the project through to completion, and also Robert Kirk and Sam Elworthy, who got it all going in the first place. The book took much longer to write than it should have—which is why I managed to get through three editors—and their patience, good advice, and immense tolerance is very much appreciated.

Thanks must also go to all the students who have taken my class on embodied cognition at the University of Lethbridge; their thoughts, ideas, comments, and suggestions have, I hope, helped make this a more accessible and interesting read than if I'd been left to my own devices. In particular, thanks to Kerri Norman, Michael Amirault, Stacey Vine, Stefanie Duguay, Clarissa Foss, Eric Stock, Kevin Mikulak, Beverley Johnson, Andy Billey, Danielle Marsh, Alena Greene, Nicole Whale-Kienzle, Joseph Vanderfluit, Amanda Smith, Brad Duce, Brett Case, Joseph MacDonald, Kevin Schenk, Mecole Maddeaux-Young, Jordon Giroux, Shand Watson, Joel Woodruff, and Ben Lowry.

I'm also very grateful to the following friends, students, and colleagues for reading and commenting on various chapters and drafts: April Takahashi, Tom Rutherford, Carling Nugent, Natalie Freeman, Graham Pasternak, Doug Vanderlaan, Shannon Digweed, Craig Roberts, and John

Lycett. I am grateful as well to the two anonymous readers who read the draft for Princeton. And I'd like to thank John Vokey and Drew Rendall for many and varied conversations on these and other topics.

Robert Barton, John Granzow, and Sergio Pellis deserve very special thanks for reading the final draft in its entirety (in John's case, more than once) and providing extensive comments. Rob and Serge played devil's advocate in the most generous way possible and helped curb my excesses, while John's enthusiasm for all things embodied and distributed ensured I didn't concede too much to the other side. All three passed on some excellent ideas, and their thoughtful, constructive criticism helped increase the clarity and precision of the arguments presented here, for which I am immensely grateful. Of course, as they say, any remaining errors are my own.

Shellie Kienzle did a superb job of proofreading the final draft for me, including pointing out those places where English idiom might raise a few North American eyebrows. At Princeton, Lauren Lepow's brilliant copyediting helped improve the text considerably, Stefani Wexler kept me on track during the final stages of manuscript preparation, and Dimitri Karetnikov drew the illustrations, based on some very poor examples and some highly convoluted explanations of what I was after. Last but certainly not least, Deanna Forrester proved herself a true friend and very generously gave up part of her Christmas holiday to construct the index. It was a genuine pleasure to work with them all.

Peter Henzi did not read a single word, but, then again, he didn't need to, as he was forced to endure an unrelenting verbal onslaught, and did so with immense good humor and grace. He also kept me in hot dinners and stiff drinks, and insisted that, sometimes, it was not only pleasant but necessary to sit on the porch in the sunshine, and I'm very pleased and grateful that he did.

BEYOND THE BRAIN

Chapter 1

Removing Ourselves from the Picture

I'm personally convinced that at least chimps do plan for future needs, that they do have this autonoetic consciousness.
—*attributed to Mathias Osvath, BBC News, March 9th 2009*[1]

I saw only Bush and it was like something black in my eyes.
—*Muntazer al-Zaidi,* Guardian, *March 13, 2009*[2]

In March 2009, a short research report in the journal *Current Biology* caught the attention of news outlets around the globe.[3] In the report, Mathias Osvath described how, over a period of ten years, Santino, a thirty-one-year-old chimpanzee living in Furuvik Zoo, Northern Sweden, would collect rocks from the bottom of the moat around his island enclosure in the morning before the zoo opened, pile them up on the side of the island visible to the public, and then spend the morning hurling his rock collection at visitors, in a highly agitated and aggressive fashion. Santino was also observed making his own missiles by dislodging pieces of concrete from the floor of his enclosure once the supply of naturally occurring rocks began to dwindle. Santino's calm, deliberate, and methodical "stockpiling" of the rocks ahead of the time they were needed was interpreted as unequivocal evidence of planning for the future.

Future planning has long been seen as a unique human trait because it is thought to require "autonoetic consciousness." Autonoetic means "self-knowing," which Osvath was quoted describing as "a consciousness that is very special, that you can close your eyes [and] you can see this inner world."[4] More precisely, it is the idea that you can understand yourself as "a self," and that you can, therefore, think about yourself in a detached fashion, considering how you might act in the future, and reflecting on what you did in the past. Osvath offered this interpretation of Santino's stockpiling behavior on the grounds that it wasn't explicable in terms of Santino's current drives or motivation, but only on the assumption that he was anticipating visitors arriving later in the day. In addition, over the ten or so years that Santino was observed behaving like this, he stockpiled the stones only during the summer months when the zoo was open. For Osvath, this spontaneous planning behavior—so reminiscent of our own—suggested that chimpanzees "probably have an 'inner world' like we have when reviewing past episodes of our lives or thinking of days to come."[5]

Of course, having a large rock flung at your paying customers by a hefty male ape is not particularly good for business, and the zoo staff were a little less impressed by Santino's antics than the scientists were. Given the suggestion that Santino possessed a highly developed form of consciousness, and an "inner world" much like our own, one might suppose that the solution to a problem like Santino would capitalize on his advanced cognitive capacities: given the ability to plan ahead and understand the consequences of his own actions—given, in other words, Santino's rationality—it would seem possible to reason with him by some means, so that he would understand why his behavior was problematic. But no. The zoo veterinarians dealt with Santino's aggressive tendencies, and so his rock-flinging antics, by castrating him.[6]

Coincidentally, the consequences of some other unwanted missile throwing were reported in the press that same week. Muntazer al-Zaidi, an Iraqi journalist, was sentenced, by a court in Baghdad, to three years in prison for throwing his shoes at President George W. Bush during a press conference held three months previously.[7] Despite the fact that al-Zaidi's actions—unlike those of Santino—were apparently not premeditated but, by his own admission, reflected his inability to control his emotions, no one (thankfully) concluded that castration would be an appropriate way to curb al-Zaidi's missile throwing. So why the difference in the chimpanzee's case?

What's Wrong with Anthropomorphism?

> Whenever you feel like criticizing someone, first walk a mile in his shoes. Then, when you do criticize that person, you'll be a mile away and you'll have his shoes.
>
> —*Anonymous*

Leaving aside the question of whether observations of one particular individual in a highly artificial setting[8] are good evidence for forward planning—let alone autonoetic consciousness—the ambivalent nature of Santino's humanlike status and the difference in response to the same behavior in chimpanzee and human is instructive. Santino's behavior was taken to indicate the presence of a "humanlike" inner life, and yet he lives under lock and key, on a moated island, his aggressive tendencies curbed by an irreversible operation. All this suggests that, despite his humanlike cognitive skills, no one expected Santino to understand why his actions were troublesome, nor did they expect him to control his behavior appropriately according to human standards of conduct. When you get right down to it, no one regarded Santino as humanlike in any way that really counted, and it remains unclear to what degree we should assume his "inner life" is anything like our own. Are we perhaps guilty of selectively "anthropomorphizing" Santino's stockpiling behavior? Are we attributing human thoughts and feelings to him simply because his behavior looks so familiar to us, and not because we really have any good evidence that he sees the world exactly as we do? Are we missing out on discovering what really makes animals like Santino tick—and what governs the behavior of many other species besides—because we're blinkered by our own human-oriented view of the world?

My answer to all those questions is yes, but let me be clear. In our everyday lives, our tendency to anthropomorphize other animals does no harm. Quite the contrary. Assuming that our dogs love us and are "happy" to see us in the exact same way that we are happy to see them can increase our sense of well-being, and it certainly benefits the dogs themselves, who are well treated and cared for as a consequence. It is also true that dogs form strong and loyal attachments to their owners, and they often pine for us when we are away from home. But none of this proves that their view of us is the same as our view of them. If we want to understand how animals work from a scientific perspective, we have to drop our anthropomorphic stance for three interrelated reasons.

Created in Our Image

First, an anthropomorphic stance means that often we end up asking scientific questions that simply reflect our own concerns. As large-brained, forward-planning, self-aware, numerate, linguistically gifted animals, we have a tendency to view each of these attributes as an unalloyed good: they serve us well, and allow us to achieve so many diverse and useful things (wheels, the printing press, combustion engines, computers, take-out pizza) that we tend to assume that similar attributes, or their precursors, would no doubt benefit other animals. So we look to see whether they have them, make our obsessions their obsessions, and see how well they measure up. This anthropocentric viewpoint fuses with our anthropomorphic tendencies so that we inevitably end up interpreting animals in human terms, regardless of whether humanlike skills would serve any real purpose for the animal in question.

The pernicious effects of such an attitude can be seen most prominently in media reports of scientific research findings. A recent report on the BBC News Web site,[9] for example, claimed that plants could "think and remember," and that they transmitted information from leaf to leaf in a manner similar to the electrical transmission that takes place in our own nervous systems. Although the report is littered with scare quotes indicating that "thought" and "memory" should perhaps not be taken literally, even a metaphorical interpretation is problematic because, as Ferris Jabr points out,[10] the analogy is far from exact[11] and creates the entirely misleading impression that plants actually do have "nervous systems" like animals, when they don't. As Jabr notes, plants are immensely sophisticated organisms that can achieve all manner of amazing things; it does them a disservice to endow them with humanlike cognitive capacities that they don't possess and don't need. Indeed, it promotes the idea that other organisms are interesting only to the degree that their capacities and abilities match our own.

An even more irritating example of this is the report of a "human-like brain found in worms"[12] (specifically, in the marine ragworm, *Platynereis dumerilii*). What this study actually shows is that ragworms possess certain cell types that correspond to those found in the brainlike structures of other invertebrates, known as "mushroom bodies," and that are also found in the mammalian cortex. In other words, the study shows that invertebrate and vertebrate brain tissue must have shared a common precursor, which evolved in the last common ancestor shared by these two groups more than six hundred million years ago. To claim that a humanlike brain has been found in worms gets the actual reasoning of the scientific article entirely backward, and generates the false impression that the whole of brain evolution has been geared toward the production of specifically humanlike brains. As my colleague John Vokey pointed out, the idea that these findings have anything to do with human brains is as ludicrous as showing that ragworms display bilateral symmetry,[13] and then declaring that the human form has been found in worms. So, again, an interesting finding, worthy of attention in its own right, gets hijacked and distorted by our strange obsession with the idea that other creatures are interesting only to the extent that they resemble us.

Whose Traits Are They Anyway?

The second reason why an anthropomorphic approach is problematic is that it cuts both ways and can create errors in both directions. While the mistaken attribution of human characteristics to other animals is the common concern, the assumption that we know exactly which traits are "uniquely human" (an assumption inherent in the concept of anthropomorphism) can result in the equivalent error of categorically denying such traits to other animals simply on the grounds that they "belong" to us.[14] The very use of the term "anthropomorphism" in this context implies that there is something very special about humans, "bursting as they are with a whole host of unique qualities that we cannot resist attributing to other beings"[15] even when they don't "deserve" it.

Both of the above problems with an anthropomorphic view spring from our overarching anthropocentrism: we consider ourselves as humans first and foremost, rather than as members of the animal kingdom, and, in so doing, we place ourselves above other animals, with the result that they inevitably fall short by comparison.[16] Consider the recent blockbuster movie *Avatar*: the gentle Na'vi of Pandora are completely in tune with nature and recognize the interconnectedness of all organisms. Nevertheless, when they entwine the nervelike tendrils of their "neural queue" (an external part of the nervous system that looks like a human hair braid) with the external neural whip of other Pandoran animals to form "Tsahaylu"—a deep bond between their nervous systems—it is the Na'vi who control the behavior of the other animals with their thoughts, and never the other way around. But why should this be? Apparently it is simply because the Na'vi are the most humanlike of all Pandora's residents (and, although the film is chock-full of exposition, this particular aspect of Na'vi life is never explained. Apparently, it is so obvious that the Na'vi should be the ones in control that it doesn't require any explanation; it is, as they say in anthropology, "an unmarked category").

This kind of anthropocentrism also means that we cannot fully appreciate our own place in nature because we are too blinkered by the traits we regard as "special." Let's consider Santino again. His behavior was taken as evidence for forward planning and the presence of autonoetic consciousness. As a result, Santino was raised up to what we clearly consider to be our own exalted level of ability, rather than leading us to question the apparent complexity of our own cognition. For if it were true that Santino possessed the ability to mentally plan his own future using a brain only one-third the size of our own, then it is equally true—and perhaps evolutionarily more valid—to argue that this ability is a general ape-level capacity and not a humanlike trait. More bluntly, it would mean that we are more mundane and apelike than we suppose, rather than that Santino is as "special" as us.

Mock Anthropomorphism, Genuine Anthropomorphism, and the Intentional Stance

Finally, anthropomorphism is a problem because the attribution of human characteristics often results in confusion about what, exactly, we have explained about an animal's behavior and psychology. More specifically, there is often confusion between so-called functional explanations that can tell us why a particular behavior evolved (why the behavior evolved in a big-picture sense; that is, how it enhances an animal's ability to survive and reproduce) and explanations of the actual "proximate" mechanisms that produce behavior in the here and now (why does the animal perform that particular behavior at that particular time?). It is perfectly reasonable to use anthropomorphic language (cautiously) in the former case as a means of generating testable hypotheses. Asking, "If I were a rat/bat/bonobo, what would I do to solve this problem?" is a useful way of going about things if we want to know why a behavior acts to increase the individual's chances of passing its genes to future generations (known as its "fitness"). This is because, as luck would have it, natural selection is a mechanism that tends to optimize behavior in exactly the way that makes this kind of intuitive sense. As John Kennedy puts it in his book *The New Anthropomorphism*: [17]

> there is no doubt that identifying the ultimate causes of any behaviour we observe is very gratifying to us. Because we are intentional beings ourselves who constantly think in such terms we long to know what an animal is 'up to', to 'make sense' of what it is doing.

Our natural tendency to assume that other people do things for a reason helps us make sense of how natural selection has acted to produce animals that behave in certain ways. It is, in other words, a metaphor: we treat the process of natural selection as though it were a person, with beliefs, desires, and plans. Anthropomorphism creeps in, however, whenever there is slippage between evolutionary explanations for behavior and explanations of the proximate physiological and psychological mechanisms that actually produce it.

For example, if we argue that male frogs sit and call by a pond all night because they "want" to attract a mate and "know" that calling will entice females, we are using the words "want" and "know" in a purely metaphorical sense. What we're really saying is that calling has been favored by natural selection because it increases the males' chances of achieving a mating relative to males that do not call. It doesn't mean that male frogs literally "know" that they "want" a mate, that they "know" they must call in order to attract one, and that they "believe" that if they call, then a female frog is sure to appear. Making any of these assumptions is anthropomorphic because we're attributing a proximate mechanism to the frogs that is, in fact, our own.[18] Evidence in support of the former statement—that calling frogs have been favored over evolutionary time because they are more successful at attracting mates—does not provide any evidence or data regarding the specific nature of the mechanisms that lead male frogs to call at ponds on any given evening. There is a clear distinction to be made between this kind of "mock anthropomorphism," which refers to the metaphorical use of anthropomorphic language in evolutionary explanations, and "genuine anthropomorphism," which refers to our tendency to assume an animal's current motivations are the same as our own without any evidence that this is the case.[19]

Mock anthropomorphism is very similar to a philosophical position known as "the intentional stance."[20] Specifically, we can predict the behavior of an organism quite accurately by treating it "as if" it possessed "intentions"—human beliefs and desires that dispose it to act in particular ways. This is because—as just noted—natural selection produces animals whose behavior we can describe metaphorically as "desiring" mates and "wanting" to call in order to attract them. Importantly, these patterns are "real" and "go all the way down,"[21] which is why we can use the intentional stance so effectively—that is, when we attribute beliefs and desires to animals in a functional evolutionary context, this isn't merely wishful thinking or naive anthropomorphism on our part, but a successful strategy that picks out a highly relevant fact about how patterns of animal behavior are organized. Our very human, highly mentalistic, take on the world allows us to predict the behavior of other animals besides ourselves because, as we noted above, it just so happens to coincide with what evolution has produced. Daniel Dennett, the philosopher who developed the idea of the "intentional stance," refers to this as "the blind, foresightless cleverness of Mother Nature, evolution, which ratified the free-floating rationale of this arrangement."[22]

It should be apparent, however, that we run into the same problem with the intentional stance that we do with mock anthropomorphism (and of course we will, since they are one and the same thing), namely, that predicting behavior isn't the same as explaining it. Indeed, Dennett explicitly recognized this problem, referring to it as "the interpretative gap"—the chasm that exists between knowing that an organism will do something in a predictable fashion and explaining why this should be the case. If we don't keep the existence of this gap in mind at all times, it becomes all too easy, and very tempting, to assume that, because we have accurately predicted an animal's behavior by attributing certain beliefs or desires to it, then we have also shown that the animal really does possess such beliefs and desires. But all we have done is named the behavior; we haven't identified or explained the mechanism by which the animal displays a particular behavior given a particular set of circumstance.[23] To do this, we need to go beyond prediction to explanation, and this requires a completely different approach.

Using Anthropomorphism Wisely?

> When we find it helpful to suppose that animals have preferences, the way we think about them is not evidence that they think.
> —*Patrick Bateson*

Given that our anthropomorphic tendencies are often useful for identifying relevant evolutionary questions and their potential answers, we needn't (and indeed, couldn't) attempt to eradicate anthropomorphism completely from studies of animals.[24] What we do need to ensure is that we think more deeply about how we are asking questions, and what our use of language implies. Given that we have only human language to describe the behavior of other animals, we can't help but use anthropomorphic terms, but we can use that language very precisely so that it is clear whether we're talking about an evolutionary solution to a problem or about the actual physiological and psychological means by which the animal achieves a solution. When we read that natural selection has favored baboon females that "decide" to groom their oldest daughters because they "want" to protect their rank position, it is very easy to slip into thinking that the female baboon is consciously making that decision—that she is choosing this from a range of options that she has weighed up carefully and consciously—because that's how we make decisions, and also because female baboons sometimes do look as though that's what they're doing, if you spend long enough watching them. But if we have only the behavior and its fitness consequences, then all we can really say is that evolution has favored females who act in this way because they tend to leave more descendants than females who don't pursue this strategy. We simply don't know what goes on in the female's head. Our own "folk psychology" that we have used to make predictions about the female's behavior doesn't allow us to assume that we also understand something about the "folk psychology" by which one baboon understands another. If we want to find out why a female baboon grooms another right now—as opposed to why grooming evolved or why it enhances fitness—we require a different set of questions and a different means of probing the animal's behavior to find the answer.

Prediction, Explanation, and Parsimony

If I seem to be laboring a point here, it is only because many critics of this kind of supposed "antianthropomorphic" position tend to attribute a more extreme view to its proponents than they actually present. A standard response to the arguments made by prominent critics of animal mental abilities[25] is to state that any opposition to attributing thoughts and feelings to other animals besides ourselves necessarily implies that animals are then treated as mindless automatons, robots, empty boxes, mere stimulus-response machines.[26] Having set up this straw man, they can then knock it down easily by expressing their incredulity that any reasonable person could hold such a position: Why do they want to deny animals these abilities? Are they perhaps afraid of their own animal natures? Are they arrogantly assuming that we are superior to the rest of the animal kingdom? The next move is to shift the burden of proof onto those who deny the similarity between our minds and those of animals. As Frans de Waal puts it:[27]

> As soon as we admit that animals are not machines, that they are more like us than automations [*sic*], then anthropodenial [the a priori rejection of shared characteristics between humans and animals] becomes impossible and anthropomorphism inevitable. Nor is anthropomorphism necessarily unscientific.

But this simply is a caricature of the actual arguments that can be made against a strongly anthropomorphic stance. First of all, to contrast "cognitive processes" with "noncognitive" stimulus-response machines or "automatons" is to generate a false dichotomy. This is because, broadly speaking, any process by which sensory input is transformed into behavioral output can be considered "cognitive," and so a stimulus-response mechanism is a legitimate "cognitive" process.[28] The alternative to assuming that animals think and feel as we do is not to assume that they have no cognitive processes at all. Rather, the more reasonable assumption is that an organism's behavior is driven by physiological processes (which include cognitive/psychological processes) that reflect the kind of nervous system it possesses, which in turn reflects the kind of body it has, which in turn is influenced by the kind of ecological niche it occupies. If we accept this as a reasonable proposition, then anthropomorphism becomes inappropriate, not because of the adoption of a some warped moral position or anthropocentric arrogance, but because other animals have different bodies and different nervous systems, and live in different habitats. This means that, even though their behavior may look similar to ours in some way or another, it need not be produced by the same underlying mechanisms (and these need not be psychological, as we'll see).

Equally, if we attempt to understand other animals only by formulating our hypotheses in anthropomorphic terms, and ask only, "What would I do if I if I were a cat, bat, or bear?" we may fail to generate a sufficiently broad range of hypotheses to test because our cultural behaviors and moral codes will impinge on this process. Consider the problem faced by a female mouse with a new litter of offspring. She needs to get enough to eat so that she can fuel milk production, while avoiding being eaten herself by a predator while out foraging. Putting oneself in a female mouse's "shoes" might well help to generate predictions that females should restrict their foraging outside the nest to times when predators aren't around, or that they should hoard food so that they can remain in the nest for longer. As the neurobiologist Mark Blumberg suggests,[29] however, it is highly unlikely that this kind of anthropomorphic projection would ever generate the hypothesis that a female should lick the anus of her young, and then eat and drink its feces and urine, as a means of regaining vital nutrients, which is the strategy that female mice actually adopt.[30]

Evolutionary and Cognitive Parsimony

One tactic used by defenders of anthropomorphism to circumvent these kinds of arguments, at least when comparing humans with other primates, is to appeal strongly to evolutionary relatedness to help them out. As Frans de Waal argues, given that humans are so closely related evolutionarily to the apes—and to chimpanzees in particular—then, if we see a behavior in a chimpanzee that looks like our own, we should be safe in assuming that similar underlying cognitive processes produce that behavior. After all, "given that a mere seven million years separate us from our chimpanzee cousins it seems most parsimonious to assume that behaviors that look the same are driven by the same processes that they are in humans."[31]

A second tactic is to make an appeal to cognitive parsimony. Cognitively simpler explanations based on "associative learning" (for example, a chain of learned stimulus-response associations between individual components of behavior) are argued to be less parsimonious than the cognitively more complex "representational" explanations they are designed to refute.[32] Here again, we should first note the false dichotomy made between "associative learning" and "cognition": associative learning is a feature of many, if not most, "cognitive" mechanisms, rather than a mechanism in itself, so asking whether something is (a) the result of "associative learning" or (b) "cognitive" is unhelpful because it is simply a confusion of logical types. Having falsely separated "associative learning" from "cognitive processes" in this way, proponents of cognitive parsimony proceed to argue that, if associative mechanisms are the only ones involved, then, to explain a complex behavior, the chain of individual associations will need to be so long, and the likelihood of all the contingent events occurring in exactly the right order will be so remote, that associative explanations actually end up being far less parsimonious than explanations that infer a more complex cognitive (and often more explicitly anthropomorphic) mechanism.[33]

Using both these tactics, de Waal argues that it is, therefore, more reasonable to accept that an animal whose behavior shows evidence of a "rich inner life"[34] is indeed possessed of one—which was exactly the argument made for the unfortunate Santino. Now, it may be more parsimonious to do this, but is this really what we're aiming for? After all, a simple explanation is not necessarily a virtue in and of itself. The real question is this: do we enhance our scientific understanding by adopting this stance? I suspect not, for the following reasons.

Why Evolutionary Parsimony Is Not a "Get Out of Jail Free" Card . . .

Let's take evolutionary parsimony first. On the one hand, given the nature of evolutionary processes, it is entirely reasonable to hypothesize that, in general, cognitive processes will be more similar between closely related species than between those that are more distantly related. On the other hand, we still have to acknowledge that this is a hypothesis to be tested; one cannot simply assume that it is true. This is because, first, our own introspection about how our own minds work need not be an accurate guide to how they actually do work. Our decision making may be much simpler than our conscious self-monitoring suggests.[35] If so, our anthropomorphizing will be doubly wrong: it assumes we understand our own cognitive mechanisms, and it then attributes this inaccurate and imperfect model to other species. A second, related problem with using behavior as a guide to our psychology is that there are many instances where behavior can be explained by more than one mechanism, and it is often very difficult to tell which one is operating in any particular instance through the observation of behavior alone.[36] The behavior seen in another species may well be consistent with the operation of some characteristically human skill—like attributing thoughts and beliefs to another individual or forward planning—but we cannot exclude an explanation based on different psychological mechanisms altogether. How, then, can we be sure that we've interpreted the behavior correctly simply by using our imperfect understanding of our own behavior as a guide? And how much more difficult is this task when we're dealing with animals that have flippers instead of hands, or can use their legs in the same way as their arms, or navigate the world using echolocation?

Another reason to be cautious about arguments based on evolutionary parsimony is that, just because, given a four-billion-year history of life on earth, seven million years is quite short, it is still a pretty long stretch of time during which significant evolutionary change is possible. Although many successful traits are, indeed, conserved across time (e.g., yeast and human beings digest sugar using exactly the same biochemical mechanisms), evolution is also a process that generates diversity. Studies on the Y chromosome of humans and chimpanzees, for example, suggest that "wholesale renovation is the paramount theme," with enormous differences in both sequence structure and gene content across the two species.[37] Given findings like these, it is prudent to consider alternative hypotheses relating to evolutionary divergence in species' capacities, even when the time spans seem relatively short. After all, just over two and half million years ago there was no such thing on the planet as a *Papio* baboon, whereas today there are at least five subspecies distributed across the whole of Africa, displaying quite distinct behavioral differences.[38] Given this, and given the divergence seen in Y chromosomes, it seems equally reasonable to suppose that, in the seven million years that separate us from the chimpanzees, there has been the potential for a significant amount of evolutionary change in both lineages, in all sorts of traits, including those relating to cognition.

Consider working memory. Chimpanzees have been shown to be far superior at tasks requiring them to reproduce a certain sequence of numbers after having them flashed briefly on a TV screen. Unlike the humans tested, the chimpanzees were both amazingly fast and impressively accurate at reproducing the correct sequence.[39] Working memory capacity in the chimpanzee line has clearly been under different selection pressures from those affecting the human lineage, so why not other kinds of psychological mechanisms as well?

. . . And Why Cognitive Parsimony Isn't One Either

This brings us to cognitive parsimony. What constitutes a "parsimonious" explanation is something of a movable feast in the field of animal psychology. In part, this reflects how researchers choose to interpret the nineteenth-century animal psychologist Conwy Lloyd Morgan's famous "canon":

> In no case may we interpret an action as the outcome of a higher psychical faculty, if it can be interpreted as the outcome of one which stands lower on the psychical scale.[40]

As such, it sounds very like the general principle known as Occam's razor, which states that "entities must not be multiplied beyond necessity"; as it is usually understood, the simplest explanation for a phenomenon is likely to be the correct one. Morgan's canon is generally assumed to be a similar "principle of parsimony" applied to animal psychology, and hence it is a recommendation that we always accept the simplest possible psychological explanation that accounts for the facts.

Interestingly, given its widespread acceptance, this interpretation is incorrect.[41] As Morgan himself made clear, simplicity should not be the measure of the correctness of an explanation because "we do not know enough about the causes of variation to be rigidly bound by the laws of parcimony [*sic*]."[42] Instead, Morgan developed his canon to make the point that, because other animals have very different sensory capacities from our own, and so encounter the world in very different ways, we should endeavor to exhaust every possible alternative explanation for their behavior before we can assume it results from psychological mechanisms similar to ours. Morgan saw no problem in attributing "higher faculties" to other animals, providing there was independent evidence to support such an attribution,[43] and he certainly didn't believe one should stick with the simplest explanation purely as a point of principle.

Despite being quite clear on these points, Morgan's canon is frequently misinterpreted as advocating exactly this kind of strictly parsimonious approach, but what then counts as parsimony is often a matter of taste. As we noted above, the convoluted nature of so-called associative learning accounts can make such explanations appear less than parsimonious, despite the simplicity of the underlying psychological mechanism. In contrast, more "cognitive" strategies, although less parsimonious in terms of the level of psychological complexity proposed, have the virtue of appearing simpler to implement and achieve than the formation of long chains of contingent associations. Accordingly, both sides of the "associative-cognitive" divide can claim parsimony as their ally on the grounds of simplicity. But the question we have to ask is this: simpler for whom? It is certainly simpler for us to understand behavior when it is described in terms that appeal to our own folk psychology, but does that really trump an argument for a simpler mechanism that happens to require a more convoluted route? Not at all. As Morgan himself noted, applying his canon anthropomorphically can mislead us:

> [B]y adopting the principle in question, we may be shutting our eyes to the simplest explanation of phenomena. Is it not simpler to explain the higher activities of animals as the direct outcome of reason or intellectual thought, than to explain them as the complex results of mere intelligence or practical sense experience? Undoubtedly, in many cases it may seem simpler. It is the apparent simplicity that leads many people to naively adopt it. But surely the simplicity of

> an explanation is no criterion of its truth. The explanation of the genesis of the organic world by direct creative fiat is far simpler than the explanation of the genesis through the indirect method of evolution.[44]

That is, although a long chain of associative responses might seem more convoluted to our eyes, and an anthropomorphic explanation much simpler, we can't decide that the latter is, in fact, more likely, based on these grounds alone. As Morgan notes, it is also much simpler to explain the existence of the organic world through the assumption that God created it in six days, than as a result of the long, slow, convoluted process of evolution, but that is no reason to accept the former explanation over the latter. As clunky and unparsimonious as it may seem, it is possible that long chains of associations are exactly the way in which many skills are learned, and complex behaviors are brought about.

Parsimony, in other words, is a red herring. It simply doesn't allow us to decide the argument one way or the other. We have to go out and test hypotheses, not simply make assumptions. If we don't test hypotheses concerning both simple and complex mechanisms rigorously and unambiguously, then we'll never know whether a more complex explanation is truly justified (especially as the power of associative learning is vastly underestimated by many of those who argue against these kind of explanations; associative learning processes have been shown to produce neural networks that can comprehend the meaning of written text, for example).[45]

A related point here is that so-called cognitive interpretations of behavior are almost inevitably pitted against the simplest possible stimulus-response forms of learning. But as we've already discussed, the opposite of a highly cognitive account of an animal's behavior is not one that posits no cognitive processing at all: first, because associative processes are themselves cognitive, but also because there are other mechanisms that can give rise to complex behavioral phenomena. As we'll see, "unfeasibly" long chains of simple associations are not the only means by which animals can produce complex and smart behavior, and it simply isn't true that the only alternative to convoluted associative chains is a heavily anthropomorphic account of psychological processes. We have to consider cognition more broadly—as the way in which animals come to know and engage with their environments, and not simply as a matter of having internal "thought processes" that are more or less similar to our own.

Minding the Gap

> The mind is like a parachute—it works only when it is open.
> —*Frank Zappa*

So those are the problems as I see them with an "anthropomorphic" approach to animal cognition and behavior. Of course, if what we want to do is ensure that chimpanzees, and other animals, are treated humanely and not exposed to potential pain and suffering, then assuming that they think exactly like us is a reasonable and defensible position. Even if they can't experience suffering, it doesn't really matter—either way, we have done no harm. In addition, it may well help us feel better as humans to treat all animals humanely, regardless of their actual capacities, so we could even be justified in doing so for these egocentric reasons alone.

If our aim is not merely to prevent potential animal suffering, however, but to understand something about the cognition and behavior of another species, then this strategy just won't do. This is not because anthropomorphism is an inherent sin, but because it does no good to merely recognize the dilemma presented by the interpretative gap, but then plow on regardless and plump for the anthropomorphic side of the argument. If we insist that other animals think and feel almost exactly as we do—or as close as makes no difference—and we can therefore explain their behavior in human terms, then we deny the animals their own voice: we impose our views on them, instead of allowing their view to be revealed to us (to the extent that we are able to appreciate it). We may do so to flatter them by allowing them human characteristics and, in so doing, flatter ourselves, but even if we don't do it for these reasons, even if our intentions are "pure" and we truly believe we are correct, the outcome is the same: we lose our ability to appreciate the animal on its own terms, and our chance to understand another way of being in the world. For scientists, the job is to jump into the interpretative gap, with all the difficulties this presents, and see how far we can explain, and not just predict, why animals do what they do when they do it.

My aim here is to attempt to shed some of the inevitable anthropocentric biases that can lead us astray, in the hope that this will help level the playing field so that we can view animal behavior and psychology—including our own—with new eyes. This simply means asking what it means to be any kind of animal, and not whether other animals are more or less like us. In some ways, then, the argument here is not really against anthropomorphism at all. The real point I'm making is that we should be wary of making premature assumptions about the level of mechanism needed to explain any particular instance of a behavior. Anthropomorphism is really just the symptom of a deeper problem: that of assuming that complex behavior and complex cognition are necessarily linked, and that only the latter can give rise to the former. All I want to do here is illustrate that this isn't the case.

Given what I've just said, it should also be clear what I won't be doing. But, just to be on the safe side, let me lay that out too: I'm not arguing that other animals do not share any of our psychological and behavioral traits, nor I am saying that only humans are clever and other animals are stupid. I am not trying to defend some kind of human superiority or specialness, and I am not writing this because I feel threatened by the notion that humans are an integral part of the animal kingdom or by the fact that we share a common evolutionary heritage with other nonhuman species. I'm not attempting a comprehensive survey of comparative psychology and cognition, and I'm not attempting to reinterpret all previous work in light of the ideas presented here, but to give no more than a brief glimpse of how some of these new approaches (and indeed some very old ones) might pay dividends. All I want to do is show how a broader perspective—one that isn't particularly concerned about whether other species possess nifty humanlike skills—allows us to appreciate other animals on their own terms, and that a reduced focus on the nature of animals' "inner lives" and greater attention to how their brains, bodies, and environments work together will give us a deeper understanding of how intelligent, adaptive behavior is produced.

Before we embark on this, however, it is worth considering the issue of anthropomorphism in a little more detail, and exploring why we are so prone to it. After all, it is well known that the first step toward changing one's views on an issue is to understand it better. A brief exploration of this topic is not only useful but also intriguing in its own right as a glimpse of an area full of fascinating research.

Chapter 2

The Anthropomorphic Animal

> You should look at certain walls stained with damp, or at stones of uneven colour . . . you will see there battles and strange figures . . . expressions of faces . . . which you will be able to reduce to their complete and proper forms.
> —*Leonardo da Vinci*

> The sea was angry that day, my friends. Like an old man trying to send back soup at a deli.
> —*George Costanza,* Seinfeld

The word "anthropomorphism" derives from the Greek *anthropos* meaning "(hu)man" and *morph* meaning "form." Originally, it was used to describe the attribution of human characteristics to the gods, but we now include other animals, inanimate objects, and even the weather[1] in the definition. The particular things that we anthropomorphize, and why, vary from culture to culture, but as many philosophers, psychologists, and anthropologists have long appreciated, all humans do it, and they do it in two distinct ways.[2]

One is to, quite literally, perceive the human form in things that are not, in fact, human. As David Hume noted, "We find human faces in the moon, armies in the clouds."[3] Not only do we perceive human faces and bodies in naturally occurring phenomena, like cloud formations, landscapes, and the like, but, across cultures and throughout history, we have deliberately captured them in pots and vessels of all kinds.[4] Today, cutlery, corkscrews, salt and pepper shakers, eggcups, fruit bowls, salad servers, oven mitts, and more are given the anthropomorphic touch. Our tendency to see "faces in the clouds" is also frequently exploited by the advertising industry. Everything from perfume bottles to fabric conditioner is produced with an anthropomorphic shape (the curvy hourglass of the female form, to be precise) and, in advertisements for alcohol in particular, the pairing of a bottle and a glass is often used to represent a man and a woman, and we perceive complex emotional relationships between them based on no more than their relative positioning (and sometimes a clever caption).[5]

Artists have recognized and depicted humanlike forms, especially faces, in natural vegetation and landscapes for as long as there's been art; one particularly nice example is that of Salvador Dalí, who, while going through a stack of photographs, initially thought he had found an unknown Picasso, a female face in semiprofile. When he looked again, however, turning the picture on its side, he discovered that it was actually a picture of an African village.[6] This example is telling because it demonstrates not only our anthropomorphic tendencies, but also how we are primed to see what we expect to see.[7] Dalí was an artist and a contemporary of Picasso; his discovery, not only of a face, but of a face painted by Picasso, reflects the background context in which he operated. The features of a Picasso were a salient stimulus for Dalí, and while we can all see the face, it's unlikely that we would all see a Picasso. The tendency of devout Christians to spot the face of Jesus in, among other things, a forkful of spaghetti, a tortilla, a chapati, and a fish finger, can perhaps be explained the same way; ditto the Virgin Mary, who has been spotted on a toasted cheese sandwich (sold, allegedly, for twenty-eight thousand dollars on eBay) and a corn kernel, and as a two-story-high image on the glass walls of a south Florida office block. The perception of a face or figure is shared by all who look at these items, but perceiving them as representations of particular individuals seems to reflect the background beliefs and concepts held by the viewer.

Animism and Anthropomorphism

> In consequence of a well-known, though inexplicable, instinctive tendency, man attributes purposes, will and causality similar to his own to all that acts and reacts around him.
> —*Théodule Ribot*

The other way that we anthropomorphize is not simply to see the human form, but to attribute to other objects thoughts, feelings, and emotions like our own. We do this with other animals, but we also do it with inanimate objects, so not only do we anthropomorphize the world, we also animate it: we perceive other objects as being alive, and we can do this without necessarily seeing them as humanlike.[8] We often talk about our cars as though they were alive, describing their performance abilities in biological, rather than mechanical, terms (they "purr like kittens"). We also do it in a less self-aware fashion: we may mistake a boulder for a bear while hiking in the Rockies. I have regularly identified, with enormous confidence, small bushes and rocks as baboons when I'm in the field and searching for a study troop, because I'm primed to expect such an encounter. Other animals seem to make similar mistakes:[9] many pet owners will have seen their cats chasing and pouncing on windblown leaves as though they were prey, and many dogs respond to sirens with their own howls. Gibbons at London Zoo do the same, responding to sirens by launching into the duets and great calls that, in their natural habitats, are used to defend their territories from other gibbon groups.

This response to the world is no accident but represents a perceptual strategy—"if in doubt, assume it's alive"—which stems from two things: the difficulty of distinguishing living things from nonliving ones, and the general difficulty of spotting live animals in their natural environments.[10] This, in turn, begins to reveal the evolutionary advantage of such a strategy: if it is hard to tell what a living thing is, and if living things are also hard to spot, then erring on the side of caution, perceptually speaking, is likely to have greater fitness benefits than assuming the reverse. Treating a large rock as though it were alive and possibly predatory is more likely to ensure one's own survival than a perceptual strategy that generally treats most things as inanimate and therefore benign.

Such a view also suggests that perception is an active process of pattern-recognition, of seeking information to fulfill certain needs and enable certain actions (like avoiding predators, or finding prey—or study animals), and not merely the passive reception of information from the environment. This point illustrates another main message of this book, which will be taken up later: namely, that to make a strong distinction between perception and cognition as separate psychological processes is both arbitrary and false.

The mistaken recognition of baboons and bears is, in most cases, easily rectified once we take a good second look, and we don't usually persist for very long in thinking that such things are alive. They fit the basic shape of a bear or baboon, but they don't act like them: even asleep, animals breathe and twitch in a way that rocks don't. There are cases, however, where such illusions persist, despite a second glance, situations where, despite knowing full well that we are dealing with the nonliving, we cannot help but see it as alive. Conversely, it is also the case that we don't make mistakes all the time and treat everything we see as being alive. What is it about certain objects that encourages persistent perceptions of animacy?

Life to the Lifeless

One of the classic studies of perceptual animacy (indeed, perhaps, of the whole of psychology) is Heider and Simmel's (1944) study of apparent motion. They showed subjects short films in which three geometric shapes (a big triangle, a small triangle, and a circle) moved around the screen in the vicinity of a large rectangle.[11] If you watch the video clip, you really can't help but see the big triangle "bullying" and "chasing" the small triangle, which is then "saved" by the circle. In the experiment, the subjects all showed the same strong tendency to describe the movements of the shapes in animate and anthropomorphic terms regardless of the instructions they were given—some going so far as to give the shapes their own distinct personality traits. Heider and Simmel suggested this came about because of the way in which the shapes moved in relation to each other.

Since these early experiments, similar studies have shown that these findings apply across different cultures[12] and ages, right down to young infants.[13] Other species of primate, besides ourselves, also seem to attribute animacy in the same manner: infant chimpanzees perceive goal directedness in geometric displays,[14] while cotton-top tamarins are sensitive to animacy, and apparently use the distinction between animate and inanimate objects to generate expectations about where objects should appear in a display.[15] A lot of effort has also gone into probing in more detail what induces the perception of animacy, and teasing this apart from the perception of more complex humanlike emotions and desires.

One particularly neat study[16] used displays containing letters as stimuli (again, about as far away from a living thing as one can get). The displays contained randomly moving letters that acted as distractors, as well as a target letter whose movement was designed to simulate meaningful, animate motion. The letter was shown either "stalking" like a predator, or "following" another letter as though worried about getting lost. By systematically varying the elements contained in these displays, the researchers were able to narrow down the features that created the impression of animacy. After viewing the displays, people were asked which one of the letters had seemed different from the others, the degree to which it seemed to be doing something purposeful, how much it interacted with other letters, and the extent to which it seemed like a living creature. As with the findings of Heider and Simmel (1944), the perception of animacy stemmed from the degree to which the trajectory of the target's motion was matched to, and interacted with, that of the distractor letters. The trajectory of the target letter was also responsible for people's stating that the letter had a "goal" and was behaving intentionally: the perception of intention was shown to become stronger when the target moved more directly toward a distractor (its "prey" or its "mother") and when it moved faster than the distractor letters.

As in Heider and Simmel's study, the stimuli used in this experiment, although they were decidedly nonanthropomorphic in shape, were situated in quite complex environments and undertook long, complex trajectories. This makes it difficult to really come to grips with why we see them as alive: if we can't easily separate the social context of the shapes from their movement, it becomes hard to work out whether our perception is caused by the spatial relationships between objects, or by something about the moving object itself.

To tease these two apart, another study used the simplest possible stimuli you can imagine: a single white dot moving on a completely featureless dark background.[17] With this setup, it was possible to show that the perception of animacy rested on two highly detectable facets of the object's motion: its change in speed and its change in direction. For example, the dot would move in a random direction with a constant motion for a certain length of time, but then it would suddenly change direction while simultaneously speeding up. If you watch displays like this, it really does seem as though the dot has suddenly "decided" to head off in a different direction. Stimuli using rectangles produced an even stronger effect, possibly because the orientation of the rectangle also changed as it changed direction, generating a stronger sense that the object was "heading off" in a different direction. In both cases, a greater impression of animacy was created if the change in angle grew steeper (the dot turned more sharply) and if the dot's final speed increased. The overall impression created by the displays is that the dot or rectangle seems to have a goal it is trying to reach, even though no such goal or target is actually visible in the display.

These results suggest that people interpret the displays as animate because they cannot easily explain them as an example of purely inanimate motion—nothing collides with the dot, or shoves it to one side, in a way that would suggest that a simple physical process caused the dot to change direction.[18] As a result, we are forced into assuming that the movement must be generated by the dot itself: that it is alive. This is interesting because, even though people know intellectually that a dot on a computer monitor cannot be alive, they can't help but see it that way. Their intellectual assessment somehow doesn't alter their perception of what happens. At one and the same time, they do and don't believe their eyes.

One reason why this might be is that, as we noted earlier, it's part of an evolutionarily honed perceptual process that is automatically triggered by anything that moves in the ways that are characteristic of living things; we show these false positives under laboratory conditions because we have the capacity to manufacture artificial stimuli that show the right kinds of movement. These automatic processes are usually highly accurate under natural conditions, but can be easily distorted by wily psychologists and their computer graphics.[19]

There is some dispute, however, over the degree to which these responses are purely automatic. There is often a high degree of variability across people in their tendency to perceive and describe the motion as animate, suggesting that sometimes people may be making more considered assessments of the displays.[20] People may interpret and report on what they have seen in animate terms simply because it is the easiest way to describe what they see, and not because they truly perceive the objects to be alive. Other researchers have argued that the experiments lack any form of ecological validity: precisely because they don't present very realistic displays, it seems possible that these effects may be produced only in the laboratory, and that they don't reflect what happens in the real world.[21] One obvious counter to this latter objection is that our everyday experience clearly suggests otherwise: we often do perceive inanimate objects as animate in our everyday lives; a falling leaf really can look like a swooping bird. The truth of the matter is that, in the real world, living things are often hard to spot and ambiguous, precisely because they've been designed that way by natural selection: camouflage, deceptive coloration (like the eyespots on butterfly wings), and crypsis often function to make the animate seem inanimate (or not present at all). Given this, a fine-tuned mechanism designed to spot life, no matter how well disguised it is, is likely to overshoot at least occasionally.[22] All that has been done in the laboratory is to distill the essence of animacy, getting rid of all extraneous "noise," so that we can pick up on the essential elements that are needed to make us think that a small red triangle is alive.

Our tendency to attribute animacy to geometric shapes forms part of a broader tendency to distinguish the movement of biological living organisms from the movement of nonbiological, inanimate entities. We are very good at determining whether movements are being made by a living being from something as minimal as a "point-light" display.[23] Researchers produce such a display by attaching lights to various areas on the body (usually the head, shoulders, elbows, arms, hips, knees, and ankles), and then filming the person in the dark so that only the lights are visible. What is interesting is that a static point-light image is meaningless to us: we see only a random assortment of bright spots. If we view a moving point-light display, however, we easily integrate the points into the image of a person walking, running, or dancing.[24] We can even determine the person's sex[25] or emotional state.[26] We can also spot the motion of other, nonhuman animals in the same way,[27] whereas articulated nonliving objects filmed in a similar manner do not tend to produce the same effects. Given that we can, with no trouble at all, see a series of lights as a person, it hardly seems surprising that we should perceive animacy in shapes that apparently have the same kind of voluntary control over their movements.

Thoughts, Feelings, and Faces

As we noted above, thinking that something might be alive is not the same as thinking it is human. In the original Heider and Simmel studies, the triangles were not seen as merely alive, but as possessing thoughts and feelings identical to our own. This was generated by the social context of the display, the way the objects moved in relation to each other, and the fact that they actually seemed to interact: they seemed to "chase" each other, "hide" from each other, "protect" or "bully" each other. This social context is what seems to promote a feeling of more than just animacy. It should also come as no surprise that the addition of features like facial expressions, or priming subjects with emotional information, also influences the degree to which a specifically anthropomorphic interpretation is given.

Adding a face—or even an approximation of a face—to anything increases our tendency to anthropomorphize. As we noted above, we often see faces in the strangest places, and there need be only the most rudimentary suggestion of eyes, nose, and mouth for us to perceive one is there. Faces are enormously salient to us—we derive an extraordinary amount of social information from facial expressions and features—so it is not surprising that we're so ready to see faces, nor that we treat anything with a face as much more human (a vegetarian friend of mine has a rule for what she can eat, which she expresses simply as "nothing with a face"; people are forever giving her carrots that look like Arnold Schwarzenegger or potatoes that look as if they're smiling, in an attempt to expose hypocrisy). Indeed, some researchers argue that there is an area of the human brain—the so-called fusiform face area (FFA)—that is highly specialized to respond to faces or facelike stimuli.[28]

Other researchers do not think that the FFA is specialized for faces in this highly specific way. Instead, they argue that it is an area that becomes specialized for stimuli that we need to recognize on an individual basis, rather than as a member of a broad category. As well as human faces, the FFA has been found to become activated in people who were expert at recognizing different kinds of cars, or different kinds of birds: this suggests that the FFA is a "flexible fusiform area," and faces just happen to be one of the most common kinds of stimuli that we need to characterize on an individual basis.[29] Our facility with faces is just one form of expertise, and one of the most highly developed, and this translates into brain areas that appear to be specialized purely for faces (and, in one sense, they are), but this doesn't mean that these same areas are not involved in processing other kinds of stimuli.

Other studies attempting to investigate this same effect suggest, however, that faces might just be a bit more special than other kinds of stimuli:

enhanced FFA processing occurs only for faces and birds (i.e., other things with faces), whereas flowers, cars, or guitars do not show this effect, nor do pictures of cars if they are shown from the side rather than from the front (i.e., in an orientation that lacks the facelike arrangement of headlights and radiator grille).[30] This also fits with research on early cataract patients. These are people born with cataracts, who then have them removed during childhood. Their inability to see early in life is a form of "natural experiment," which gives us the means to investigate how experience tunes face recognition early in life. What's interesting about people who have had early cataracts is that they are insensitive to the configuration of human faces, and they don't seem to use any kind of "holistic processing" (that is, to treat a face as a "whole," rather than as a collection of features. Holistic processing is the reason why, among other things, upside-down faces of people we know are harder to recognize than right-side-up ones).[31] What is even more interesting is that this lack of sensitivity to configuration does not extend to any other kind of object. People who have had early cataracts removed from both eyes can perform completely normally on non-face-related tasks that require them to judge the spacing in a simple pattern of five shapes, or to match shapes with each other, whereas they fail miserably at the same kinds of tasks when they involve faces.[32] Perhaps the best way to characterize the nature of brain specializations for face processing, then, is to argue that faces are likely to be processed by multiple areas in the brain (including, but not limited to, the FFA) that, although not dedicated to faces alone, are activated to a much greater extent by faces and other facelike stimuli than by other kinds of visual signals. The greater sensitivity to faces in these areas likely reflects our greater experience with individual faces and the importance of reading the signals they produce.[33] As we shall see in chapter 5, a predisposition to respond to certain kinds of features regularly found in a creature's environment, combined with learning from experience, is common to many animals.

Babies and Faces

As well as being able to see faces in all kinds of objects, we also actively prefer to look at faces over other kinds of stimuli, right from the time we're babies. It used to be thought that babies had some innate (meaning, quite literally, "present at birth") face-recognition abilities built into them: babies prefer to look at faces within only a minute of birth (that is, they spend longer looking at a facelike stimulus than at one that does not resemble a face).[34] In order to rule out any kind of rapid learning (as seen, for example, in young geese, who learn the features of their mother extremely rapidly after birth, a process known as imprinting: see chapter 5), the experimenters in these studies ensured that the babies had never seen a real, live face in the time between their birth and the test.

More recently, studies have suggested that this innate bias may not be for faces as such, but for the particular kind of geometrical configuration that faces present.[35] In these studies, babies were shown simple arrangements of shapes that had "top-bottom asymmetry," that is, they were top-heavy, with more elements present in the top half of the stimulus than in the bottom. Although there is something vaguely facelike about these displays, they don't produce the immediate recognition of a face; you have to work at it a little. The babies' preference for this stimulus was compared to their preference for a stimulus that was bottom-heavy. As predicted, the babies spent more time looking at the top-down asymmetric stimulus compared to the bottom-up one.

This effect persisted when real faces were used in which the facial elements were scrambled and therefore in the wrong place: as well as preferring an upright top-heavy face to a bottom-heavy upside-down one (achieved through the reversal of the facial elements only; that is, an upside-down face was presented in an upright head), babies also preferred a completely scrambled face that was top-heavy (one that had all the elements of a face, but placed in odd positions) to one that was bottom-heavy. What they apparently didn't distinguish between, however, was a totally normal upright face and a completely scrambled top-heavy face: babies looked at both kinds of faces for equivalent amounts of time, despite the fact that the scrambled top-heavy face looks like a particularly strange Picasso portrait. This suggests that babies are not born with an ability to recognize or prefer faces at all, but they simply show a bias for any stimulus that has a top-heavy spatial arrangement of elements. The early experiments, with their highly facelike stimuli, couldn't distinguish this bias toward top-down asymmetry from a bias toward actual faces.

This general preference for top-heavy arrangements becomes fine-tuned to become a specific preference for faces during the early weeks and months of life as a result of experience. Natural selection seems to

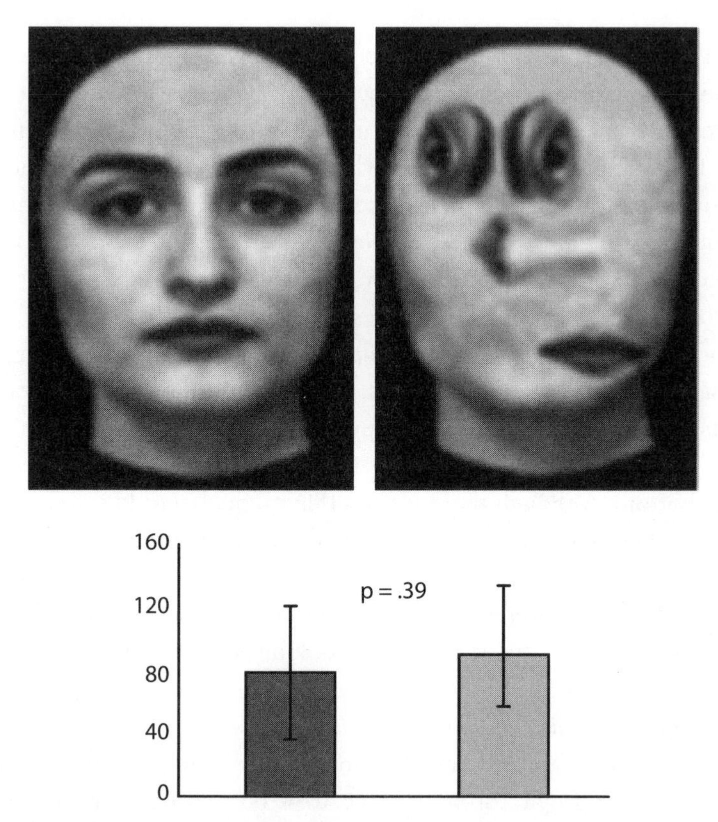

Figure 2.1. Babies do not differentiate between a normal image of a face and an image with all the features scrambled. The slight blurring of the images is not a problem in studies of this nature, as young babies' vision is very poorly developed. (Looking time, in seconds, is graphed on the *y* axis.) Redrawn with permission from Sage Publications.

have provided us with some very basic perceptual constraints on our visual processing abilities that are "experience-expectant": that is, they require exposure to faces in order to narrow down the category of stimuli to which they respond. Given that a human infant will inevitably encounter a human face within moments of birth (devious scientists aside), it is much more cost-effective, from an evolutionary perspective, for an organism to develop with only a very basic face-recognition mechanism and then let all the faces in the environment do the work of refining it, than it is to develop a brain with a highly specialized and fully functional face-recognition mechanism at birth. The latter would require much more brain tissue, for a start, which could lead to problems at birth: babies are born when their brains are still quite small so that their heads can fit through the mother's pelvis and birth canal, and most of human brain growth takes place during the first year of life, once we're safely out in the open. A more highly developed face-recognition mechanism could increase brain size to an extent that could make birth more difficult.

A very basic "experience-expectant" bias is also a more flexible mechanism than a fully preprogrammed recognition device; babies can then learn to recognize the individually distinctive kinds of faces they encounter as they develop, whereas a preprogrammed mechanism runs the risk that, if faces seen didn't quite fit the specific face-template developed before birth, then babies wouldn't be able to recognize faces at all. The simple bias mechanism allows human infants to lock onto the facial differences that are important, to family and population facial traits, as well as to the facial traits common to all humans.

Our exceptional ability to recognize faces arises out of a very general mechanism that is shaped by the experiences to which we are inevitably exposed. So faces both are and aren't special: any object with the right kind of top-heavy bias is equally fascinating to us when we are very tiny. It is just that an object with paired dark horizontal blobs in its upper parts is most likely to be a face, and the constant, intense, and rewarding exposure that we receive from faces means that, compared to other objects, faces acquire a prominence and significance for us that few other objects can match. This is a point worth noting because it shows how fundamentally an organism—in this case a human infant—is embedded in, and inseparable from, its environment. This is something we'll be returning to at various points in what follows.

As well as faces, other humanlike body parts increase the likelihood that we will view something as human. The Italian designers Alessi have got this down to an extremely fine art: their tea strainers, corkscrews, and eggcups are all highly anthropomorphic, and, while we don't necessarily impute thoughts and feelings to them, they may well increase our willingness to touch them and use them, and, of course, the main aim of the exercise, part with our cash so we can furnish our kitchens with them. This much is obvious: the more something looks like us, the more likely we are to treat it as though it is one of us. As the old saying goes, if it looks like a duck, and quacks like a duck, then it is a duck. As we discussed earlier, this isn't always true: it may quack, but, in reality, there may be no duck there at all. We can be fooled, and this means that, if we're not careful, we can make mistakes.

Selection for Sociability

> It is good to rub and polish our brain against that of others.
> —*Montaigne*

So our perceptual systems seem geared to recognize animate beings, and we then endow these beings with goals and intentions (even if they aren't animate at all), especially if they are engaged in social interaction. As suggested above, this may reflect the need to be able to detect potentially dangerous forms of life quickly, but it also reflects our current status, and past evolutionary history, as intensely social primates. To live in a social group, and to be good at it, means possessing the ability to read the signals given by other animals, and to respond to them appropriately.

Over evolutionary time, the primate brain has undergone changes in both size and structure, which seem to reflect the demands of living in social groups.[36] Leslie Brothers, a neuroscientist turned psychoanalyst, was among the first to identify a number of primate brain regions, namely, the amygdala, medial temporal lobe, orbitofrontal cortex, and superior temporal gyrus, that were involved in responses to social stimuli produced by another individual; she dubbed this the "social brain."[37] As well as having regions specialized for the processing of social interactions, primates also show an overall increase in brain size compared to other mammals. In particular, the part of the brain known as the neocortex (the most recently evolved part of the brain) has undergone a quite dramatic expansion.[38] The size of the neocortex is related to the size of the group in which a species lives: primates living in larger groups tend to have larger brains. This finding has been taken as evidence in favor of the "Machiavellian," or social intelligence hypothesis,[39] which argues that life in a structured social group (one composed of different generations of animals of varying degrees of relatedness) is inherently complex, and its members require more highly sophisticated cognitive mechanisms in order to cooperate and compete effectively within such a group.[40]

More interesting, however, from our point of view, are detailed studies by psychologist Robert Barton,[41] in which he aimed to identify exactly which parts of the neocortex had expanded over evolutionary time. Barton has shown that the visual cortex (in particular, an area known as V1) has undergone the greatest expansion across the primates, an expansion that is associated with both fruit eating and a switch to being active during daylight hours (diurnal).[42]

It is easy to see why the visual cortex should be larger among diurnal primates, since these are much more reliant on vision for getting around than are their nocturnal cousins that rely on smell. Equally, fruit varies in color according to ripeness, and this has to be spotted amid a background of green, so color vision is a great benefit to fruit-eaters, making it obvious why this would also select for larger visual areas. In addition, Barton found an intriguing link between particular features of the visual system and social group size. This is a little less obvious to work out: why should bigger groups require better vision? Understanding this requires us to delve a little deeper into some primate brain anatomy.

In the primate visual system there are two distinct pathways by which visual information is transmitted from the visual areas at the rear of the brain to the frontal lobes where it is processed further.[43] One, known as the magnocellular pathway, is concerned with movement detection and is common among all the mammals, while the parvocellular pathway is unique to the primates and is associated with the detection of fine detail and color. The magnocellular pathway passes over the dorsal (top) area of the brain, while the parvocellular pathway follows a ventral (bottom) route and is linked to the amygdala, which is involved in the perception and processing of emotions (and part of Brothers's "social brain").

Among diurnal primates, it is the brain cells associated with the parvocellular pathway specifically that are positively related to social group size, whereas the magnocellular pathway is not, suggesting that it is the analysis of fine detail with color that is socially relevant. Barton argues that the parvocellular pathway was enhanced during primate evolution in order to process particularly important details of dynamic social stimuli, like facial expressions, gaze direction, and posture.

Monkeys and apes (and humans) all communicate using facial expressions and postural displays that indicate whether they are threatening another, appeasing another, or encouraging mating. These facial expressions and postures are often enhanced by colorful skin or striking fur color, making it easier to project and pick up signals over a distance. As well recognizing these signals, monkeys and apes also can produce an appropriate response to them. It is here that the parvocellular pathway's link with the amygdala becomes relevant. The amygdala is involved in giving emotional valence (i.e., its intrinsic attractiveness or aversiveness) to a particular signal.[44] For monkeys, recognizing whether another animal is threatening ("anger") or showing submission ("fear") tells them a lot about their standing in the group, and what their next move should be, and these are the emotions most frequently linked to the amygdala. The expansion of the neocortex in primates to a large extent reflects an increased need for better recognition and interpretation of the visual signs and signals given by the other animals in the group, from which are built the many and varied social interactions that monkeys and apes engage in each day.

The Social Life of the Brain

More detailed work by the neurophysiologist and psychologist David Perrett and his colleagues has also revealed how the social world of the primates has shaped the brain.[45] His studies have revealed that an area of the brain called the anterior superior temporal sulcus, (STSa) (through which the parvocellular pathway runs) contains brain cells (neurons) that show highly specific responses to particular kinds of social stimuli. For example, there are neurons that respond only to a head facing to the left, but not to one in any other orientation. There are also neurons that respond purely to faces, facial expression, and eye gaze direction, and similar neurons are found in the amygdala. For gaze direction, in particular, there are several populations of neurons that respond to different gaze directions, with one of the most significant representing neurons that respond to direct eye gaze toward the subject. This makes sense because, as well as detecting whether another animal is using a threatening facial expression, an individual also benefits greatly from being able to detect whether such a signal is directed toward it (in which case direct action is needed) or is focused on another individual (in which case, there may be no immediate danger). There are also neurons that exclusively respond to biological motion, helping to explain why point-light displays are so easily recognized. The primate brain is highly specialized to help us recognize particular social stimuli, like the body movement, faces, and expressions of other individuals, and to produce an emotional response toward them.

The recent discovery of "mirror neurons" makes this point even more forcefully.[46] These are neurons, first discovered in the premotor cortex of macaque monkeys, that fire both when an animal performs an action (such a grasping a food item) and also when the animal simply observes another individual performing that same action. In the latter case, strong inhibitory signals prevent neuronal impulses from reaching the limbs, so that brain activity doesn't translate into actual movement. Mirror systems have also been identified in the human brain.[47] In both monkeys and humans, these systems have been found not only for motor actions, but also for communicative gestures, such as lip smacking in monkeys (an affiliative gesture) and expressions of disgust in humans.[48]

Mirror neurons have quite rightly been hailed as an important discovery for all kinds of reasons, but the one most relevant here is that they again illustrate the immense importance of social signals to primates: actions produced by mechanical objects but otherwise identical to those performed by a human or a monkey do not trigger a mirror response, only those from another animal. This resembles the actions of the neurons identified in the STS and amygdala, but what's even more interesting here is that the "mirroring" response of the neurons means that, at a very fundamental, strictly nonconscious level, an animal with a mirror system will find the actions performed by others significant and meaningful because it is, in a sense, performing the actions itself.[49] Mirror neurons therefore show how action and perception are part and parcel of the same process, rather than distinct from each other. In other words—and to introduce a phrase that will be bandied about frequently in this book—psychological processes are "embodied": they are not things that somehow "float free" from the animal, but are firmly grounded in the physical actions of the animal's body both as it observes other animals, and, of course, as it moves around the world itself.[50]

A more recent study[51] by Robert Barton on the correlated evolution of the neocortex and the cerebellum[52] makes a similar point: this shows that there are particular systems and networks in these two areas that have been selected over evolutionary time (rather than the simple expansion of those parts that we tend to associate—rather anthropocentrically—with complex "thought"), and that these patterns of correlated "mosaic" evolution suggest strongly that brain function is geared to coordinating perception and action in a dynamic environment.

These kinds of basic neural adaptations, and their often highly visual nature, make sense evolutionarily because, due to the lack of any kind of language among monkeys and apes, the only information these animals can obtain about others is, in the main, that which they can see with their own eyes. The "social intelligence" of the primates is, in essence, based on visually guided and emotion-based action in the world. Our tendency to see other creatures in terms of ourselves rests largely on this basic primate tendency to respond to certain forms of social stimuli.

This isn't the whole story, of course: it doesn't explain why animate motion and social engagement give rise to the attribution of specific kinds of thoughts, beliefs, and desires to triangles, beer bottles, and baboons. Instead, this clearly relates to the manner in which humans develop: over a long period, with both linguistic and cultural support, we are "trained" by other humans to see our public behavior as the reflection of private internal mental states and to then use these mental state atttributions to explain our own and others' behavior. [53] In particular, the Russian developmental psychologist Lev Vygotsky argued that we can understand our private mental functioning only through a close examination of the very public social, cultural, and historical processes that shape it. This means that we cannot make any firm distinction between an individual's world and her social world because they are so closely intertwined with each other that they are, for all intents and purposes, the same: we come to possess our particular ways of behaving, with our particular attitudes, thoughts, beliefs, and knowledge of the world, only because, from the very beginning of our lives, we have been engaged in social processes that involve other individuals embedded within, and permeated by, our various cultural practices.[54]

So, as Vygotsky conceived of it, a child's mental processes are not the source and cause of her behavior in the world; rather a child's behavior in the world is the source and cause of what eventually ends up in her head: the exact reverse of what most modern psychology would have us think. What children learn over the course of development, then, is how to participate in the shared practices and rituals of their culture, and they learn this through their own active participation, supported by more knowledgeable adults. And part of what children learn—particularly in Western society—is to explain people's behavior in terms of beliefs and desires that are invisible to or hidden from other people. This isn't to say that we truly have some kind of deep insight into another person's mind, or that we possess a capacity to tap into another's brain states in some way, as some researchers have argued. It's much more likely that we respond to a whole package of observable features, like a person's body language, his actions, what he says, and the contexts in which these occur, which we then label as his "thoughts" or "beliefs" about the world.

Now, given this lifelong immersion in a human world where we use invisible "thoughts" and "beliefs" to explain observable behavior, it is unsurprising that the more similar another creature, whether real or artificial, is to us, and the more likely it is to produce social signals of the kind most salient to us, the more likely we are to assume that we can use the same reasoning and mental state attributions that we use on each other to account for what is going on inside its head.

This evolutionary "priming," and its development into a compulsive anthropomorphic tendency that we don't even notice ourselves using most of the time, means that we have to unlearn many of the things we currently take for granted, and realize that seeing ourselves in other animals may sometimes be mistaken. The upside of this, however, is that understanding a bit more about how our anthropomorphism comes about allows us to recognize that, as Darwin suggested, there is indeed continuity between species, and that we share many traits with our primate and, more generally, our mammalian cousins.[55] In other words, there will be times when there is no possibility of anthropomorphizing because other animals really do share some of our characteristics. As we noted in the previous chapter, we do, however, need to be careful to keep our primate social biases in check, so that we don't fall prey to projecting our own emotions, intentions, and motives onto other animals. Remember, we are easily fooled; a sad-looking triangle doesn't really feel sad.

So that's one of our primate biases dealt with, but as we now launch into an exploration of how other animals deal with the world in a nonhuman, nonprimate manner, we will have to confront another of our inherent prejudices toward other animals: namely, the big-headed bias that comes with having an enormously large brain.

Chapter 3

Small Brains, Smart Behavior

So this gentleman said a girl with brains ought to do something else with them besides think.
—*Anita Loos,* Gentlemen Prefer Blondes

Big brains are our defining feature as humans. No other animal has one quite as big relative to its body size, and we rightly attribute our current domination of the planet to the kinds of advanced cognition and flexible behavior that our big brains make possible. A big brain is a good brain then, as far as we are concerned, which, as you should now recognize, is an essentially anthropocentric view. No matter how often we are told that humans are not the pinnacle of evolutionary achievement; that evolution is a bush, not a tree or a ladder;[1] that there is no "great chain of being" leading from lower to higher life-forms, we still feel that our big brains, and the sophisticated cognitive abilities and behavioral flexibility associated with them, represent a significant evolutionary advance. Even if we don't state it so baldly, we inevitably end up concluding that we are superior to animals that have only very small brains.

In some senses, of course, this is true: we are capable of feats that no other animal has ever achieved, from inventing the wheel to performing heart transplants to sending rockets to the moon. On the other hand, does this really mean that humans are superior to, or more successful than, other species, or are we merely different? If one takes another measure of superiority, humans don't appear so impressive. For example, it is estimated that the average number of insects found per square kilometer equals the total number of people on the earth. Insects must be doing something right. Why, then, do we assume that big brains are inevitably better than small ones?

Part of the reason is that our definitions of success and superiority are obviously very narrow and human oriented. Another, perhaps more important, part of the reason is that, from our human-biased perspective, we tend to focus only on the benefits of a large brain and fail to consider the costs, which are substantial. Our brains are immensely expensive organs to maintain: despite constituting only 2% of our body weight, they use up fully 20% of our total energy requirements, just to run idle. They are prone to damage and need to be kept within a fairly circumscribed temperature range to stop them from scrambling; they take years to develop and mature; and our children are born so helpless that they cannot take care of themselves at all, and they do not reach independence and sexual maturity until an age when most mammals have already grown up and reproduced, and are well on the way toward advanced old age, if they haven't already died. To balance these heavy costs, and so be favored by natural selection, big brains must confer an equally large benefit.

When one starts considering what this might be, however, the answer isn't so obvious as one might think, since many animals display a remarkable level of behavioral complexity and flexibility, all with brains the size of a pinhead. Take face recognition, for instance, which we considered in the previous chapter. The skill with which we recognize faces, and distinguish between individuals, has led to the suggestion that this is a specialized human skill, requiring dedicated brain areas and a lot of neuronal processing. Just like humans, however, paper wasps (*Polistes* spp.) are able to recognize the facial features of other individuals and use this ability to help maintain a strong social order in the hive.[2] In an ingenious experiment, in which the markings on wasps' "faces" were changed experimentally (researchers painted additional yellow marks or covered up existing ones), wasps no longer recognized altered individuals and directed more aggression toward them. Aggression is the means by which wasps test each other and establish relative dominance. The increase in aggression observed toward experimentally altered wasps relative to control wasps, which had not had their markings changed, suggests that the wasps perceived their nestmates as strangers and needed to establish rank with these "new" wasps.

An alternative explanation is that the facial markings do not provide information on individual identity, but rather provide a more general kind of information regarding dominance status or role within the colony. Facial markings would then function more as a uniform does in humans: to mark someone out as fulfilling a particular role, like a nurse or a policeman. The level of aggression shown to the experimental wasps diminished over time, however; if the changes in markings merely signaled a change in a wasp's status, then one would expect the new level of aggression to be sustained because the aggressive response would be to the new role and not the "new" wasp. As aggression decreased, it seems that the wasps were responding to the "new" wasp as an individual, and gradually learning to associate that individual's rank with its particular facial markings. It seems that wasps' "faces" really do provide information on individual identity, and that the wasps use this information to regulate their social interactions with each other.

Honeybees, meanwhile, can discriminate between human faces, and learn to associate a particular face with the provision of a reward. They can also distinguish a rewarded target face from a bunch of novel distractor faces. Even more interestingly, honeybees find it more difficult to recognize familiar human faces if they are displayed upside down, suggesting that, like humans, they may use some form of configural processing to recognize faces (i.e., they take into account the relative positioning of the features, as well as their individual size and shape, and they process the face as a whole, not as a collection of parts).[3] Of course, honeybees don't actually recognize human faces as human faces, nor do they attach any significance to them other than as cues to food, but this example, as well as that of the *Polistes* wasps, suggests that we need to keep an open mind about both the uniqueness of our own abilities and the potential for complex behavioral responses emerging from nervous systems that are much simpler than our own.

In this chapter, we explore the worlds of so-called simpler animals, both real and robotic, so that we can more fully appreciate how much a small brain can achieve.

The Parable of the Ant

> It is not enough to be busy. So are the ants. The question is: what are we busy about?
>
> —*Henry David Thoreau*

Our irresistible tendency to see things in human terms—that we are often mistaken in attributing complex human motives and processing abilities to other species—does not mean that an animal's behavior is not, in fact, complex. Rather, it means that the complexity of the animal's behavior is not purely a product of its internal complexity. Herbert Simons's "parable of the ant" makes this point very vividly.[4] Imagine an ant walking along a beach, and visualize tracking the trajectory of the ant as it moves. The trajectory would show a lot of twists and turns, and would be very irregular and complicated. One could then suppose that the ant had equally complicated internal navigational abilities, and work out what these were likely to be by analyzing the trajectory to infer the rules and mechanisms that could produce such a complex navigational path. The complexity of the trajectory, however, "is really a complexity in the surface of the beach, not a complexity in the ant."[5] In reality, the ant may be using a set of very simple rules: it is the interaction of these rules with the environment that actually produces the complex trajectory, not the ant alone.

Put more generally, the parable of the ant illustrates that there is no necessary correlation between the complexity of an observed behavior and the complexity of the mechanism that produces it. Another illustration of this comes from work on autonomous, mobile robots. These are built to move around and engage with the world physically, providing researchers with the means to investigate precisely these kinds of synergistic interactions between organism and environment (see also chapter 5). As a result, this area of research is also known as "behavior-based robotics." By building systems where the internal complexity of the animal is fully specified and known to the researcher, and then exposing them to different environments, researchers can investigate the behavior of these "creatures" and, more importantly, understand and explain them.

The "Fertile Turtles"

The pioneer of behavior-based robotics was William Grey Walter, a neurophysiologist by training,[6] and something of a character: as well as a robotics pioneer, he was also a "home explosives expert, wife swapper, t.v. pundit, experimental drugs user and skin diver."[7] In the 1940s, he created two autonomous robots, which he called Elsie and Elmer (drawing on the terms used to describe them, "Electro Mechanical Robots, Light Sensitive, with Internal and External Stability").[8] These were small, motorized tricycles with a transparent shell over the top, which made them look rather like tortoises (or turtles, if you live in the United States). His reason for building them was to test his idea that the complexity of brain functions arises not necessarily from the number of neurons but from the "richness of their interconnections,"[9] and to work out how different patterns of interconnectedness actually function to produce different kinds of behavior.

The tortoises were built to make this problem manageable. Each had only two "brain cells" that could produce one reflex response each, and each was built from some very simple components (two small motors, two batteries, some vacuum tubes, two condensers, and two relays): one reflex dealt with the tortoises' response to light, the other with their response to touch. How many ways of behaving would be possible for a creature like this—a creature that possessed only two "brain cells"?[10]

The light reflex was under the control of a photoelectric cell (an "eye") mounted on the front of the steering column of the tricycle, which always pointed in the same direction as the front wheel. The eye was covered with a hood that blocked out light from all directions except the front. The tortoise also had a headlamp—a "running light"—that lit up when the robot was moving so that Grey Walter could tell it was working correctly. The touch sensor was an electrical contact that closed whenever the "shell" of the tortoise came up against an obstacle. The tortoises were wired in such a way that they would seek out light sources and attempt to overcome obstacles, by either knocking them out of the way or moving around them.

Seeing the Light

If the tortoise was switched on in the dark—or under very low light levels that the "eye" couldn't detect—the driving motor turning the front wheel would operate at half speed, propelling the robot forward, while the motor on the steering column itself would rotate the front wheel at full speed. The eye would also rotate, because it was attached to the rotating steering column, so that it always "looked" in the direction in which the tortoise was moving. This allowed the tortoise to "scan" its environment. The back wheels were not powered but ran free, so the tortoises were fully front-wheel drive vehicles. The combination of the driving motor propelling the tortoise forward and the circular rotation of the front "steering-scanning" wheel meant that the tortoise slowly moved in a series of looping arches, "exploring" its environment.

If, during this exploration, a light source was detected by the "eye," the steering motor would stop, and the steering-scanning wheel would therefore stop rotating, becoming fixed at the angle at which it had been turned when the light was detected. The headlamp would turn off as a result, and this had the effect of turning up the driving motor to full speed. The tortoise would stop "scanning" and would begin to "hurry" toward the light. If the eye was at an angle when the light source was detected, rather than pointing straight ahead, then this "hurrying" movement would actually begin to deflect the tortoise away from the light. The result would be that the activity in the eye (the photoelectric cell) would fall below its threshold for activation, so that the front wheel would begin rotating, and the robot would again begin its slow scanning. This would soon result in the tortoise's once again detecting the light source, triggering another bout of hurrying behavior. In this way, the tortoise would approach the light by a series of successive approximations, each one more accurate than the last, because the aiming error was reduced each time the tortoise got nearer to its goal.

As the tortoise got closer to the light, it would "see" the light getting brighter and brighter. If the light was a very bright one (say, a forty-watt lamp or a flashlight), then, when the tortoise got very close, its behavior pattern would change again to "dazzle" mode, and it would then avoid the light: the front steering-scanning wheel would be activated at half speed, while the driving motor would keep going at full speed so that, instead of reverting to its slow scanning behavior, the "dazzled" tortoise would smoothly veer away from the light before it bumped into it (thus avoiding the fate of many moths . . .). Grey Walter suggested that the tortoises' behavior was like that of animal seeking an "optimum" source of light—one that was not too bright and not too dim. Once it reached such a light source, it would circle around it, whereas it would leave lights that were too bright and continue to explore for something more suitable. The light reflex is therefore an example of a negative feedback process, like a thermostat that controls room temperature: bright light caused the tortoise to react in a way that reduced the amount of light detected, while low light caused the robot to respond in a way that increased the amount of light detected, so that, overall, a stable level of light stimulation was achieved.

The light reflex also allowed for some intriguing "emergent" behaviors. For example, although the headlamp was there only to indicate that the tortoises were running correctly, it led to some very interesting behavior when two tortoises were placed together. As described above, the tortoises would wander around, scanning, and would soon detect each other's headlamps. The detection of the headlamp light on the other tortoise would switch them to "hurry" mode, but, of course, each achieved this by switching off its own headlamp. All of a sudden, then, there was no light source for either tortoise to approach. With no light to be attracted to, the tortoises would switch back to scanning mode, and the headlamps would come on again, and the same interaction seen previously would ensue: the tortoises would be attracted to each other initially but would be brought up short by the sudden disappearance of the light to which they had been attracted. As Grey Walter suggested, this looked very much like a "mating dance"—one in which "the machines cannot escape from one another; but nor can they ever communicate their 'desire'"—but this behavior was not programmed into the tortoises; it simply emerged from the way they were set up and their interaction with the environment.[11] Similarly, when the tortoises were displayed at a meeting of the British Association for the Advancement of Science, they manifested an emergent and unexpected penchant for women's legs, presumably because they were attracted by the light-reflecting properties of the women's nylon stockings.[12]

This sensitivity to light allowed Grey Walter to devise a means by which the tortoises could "feed," or, more accurately, recharge their own batteries. A hutch was built containing a bright light. When the tortoises' batteries were fully charged, the light would set the robots into dazzle mode and so repel them. As the batteries ran down, however, the tortoises would be increasingly attracted to the light because they would sense the light as being of lower intensity. This attraction to the light source would eventually cause the tortoises to move right into the hutch; once they were there, the batteries would be recharged. Once a tortoise was recharged, of course, its sensor would register the true intensity of the light, and the brightness would trigger the dazzle response. The tortoise would leave the hutch, repelled by the bright light, until it was "hungry" again. There are, however, no records showing that this recharging behavior was 100%

successful,[13] although the tortoises clearly could move in and out of the hutch using their light-seeking reflex.

Avoiding Obstacles

The touch reflex was less sophisticated than the tortoises' light sensitivity but still allowed for some fairly sophisticated behavior to emerge. It was wired up so that, when the contact was closed, the signal coming from the light sensor was "ignored," because the vacuum tubes that usually acted as amplifiers of the light signal would instead become part of an oscillating signal that caused both the driving and steering motors to run at full power, switch off, and then switch on again. What this meant behaviorally was that, with an obstacle in the way, the tortoise would ignore all other stimuli, and would push and shove the obstacle, bouncing off it and then approaching it again, until it had either pushed it out of the way or managed to move past it. The touch reflex is therefore an example of a positive feedback process; it doesn't maintain stability, like the light reflex, but keeps the tortoise pushing harder and harder for clearance as it continues to encounter an obstacle. As the tortoise wandered about its environment, exploring, detecting and moving toward light sources, it would also push any obstacles out of the way so that, on occasion, the obstacles would end up piled neatly against the walls (which is another interesting form of emergent behavior that we'll deal with in more detail below).

Grey Walter suggested that these reflexes could also work in concert to produce remarkably flexible and lifelike behavior: when Elsie was confronted with a mirror, for example, it would approach it because it would detect its own headlamp on the front of its shell. Detection of the light would, of course, immediately extinguish the headlamp as the tortoise switched to hurry mode. With no light to be detected, the robot would immediately switch back to scanning mode, the headlamp would come on again, and the tortoise would again "see" its own headlamp's reflection and switch to hurry mode. This turned off the headlamp, switched the robot back to explore mode, and . . . so on and so on. In other words, a feedback circuit would be set up, in which the environment—the mirror—played a crucial role.[14] The robot would appear to be "dancing" in front of the mirror, which, as Grey Walter himself pointed out, "on a purely empirical basis, if it were observed in an animal, might be accepted as evidence of some degree of self-awareness."[15] Grey Walter referred to his robots as the "fertile turtles" precisely because their complex and unpredictable behavior generated so much discussion about how real animals worked.[16]

Grey Walter's robots highlight perfectly how we cannot simply translate behavioral complexity into an assessment of an animal's cognitive complexity. Without a knowledge of how Elsie and Elmer were put together, it would be natural to attribute more internal complexity than was warranted, as well as to overestimate the cognitive demands of the tasks that they could perform, like "recognizing" themselves in a mirror. Our tendency to focus only on what's going on in an animal's head, when we seek to understand how and why it behaves, means that we fail to notice the extent to which the structure of the environment and the physical shape of the animal's body play a highly active role in shaping its behavior.

This was brought home to me in another way, when one of our undergraduate students, Tom Rutherford, asked if he could attempt to replicate Grey Walter's robots using a Lego Mindstorms robot kit. This proved to be harder than it looked because of the constraints on how Lego robots are put together (for example, a power cable connects the robot's "brain" to the drive motor, so that it isn't possible for the robots to steer continuously in one direction as Grey Walter's robots did, and the drive wheel and light sensor cannot rotate 360 degrees but can only sweep from left to right in an arc) and also because the software programming of the robot's brain proved to be far more complex than Grey Walter's simple vacuum tube system. Behaviorally, while Tom's Lego robot could detect light successfully and begin to move toward it, the difference in its body compared with the original tortoise robot meant that it couldn't straighten itself out quickly enough to continue in the direction of the light, but would sometimes turn its back on the light completely and so head off exploring in another direction. With characteristic understatement, Tom noted, in the term paper he wrote on his robot-building exploits: "It is with considerable effort that modern software and hardware are able to imitate the simple brain of the original turtle. . . . It seems that Grey Walter's simple, specifically designed circuit was able to produce more accurate behavior than modern robotics." Despite having a better "brain," Tom's robot proved less effective than Grey Walter's tortoises because of the way in which its body encountered the environment. We can use another—more recent—robotic example to illustrate this point even more vividly.[17]

The "Swiss Robots"

A didabot is a small, wheeled robot. If a number of these are placed into an arena in which small cubes are scattered randomly, they move around clustering the cubes together, so that eventually there are only two large clusters of cubes, with a few cubes left here and there against the walls. The robots are dubbed "Swiss," because, like their namesakes, they are incredibly neat and tidy, and their aim, apparently, is to clear the arena of clutter. As with Elsie's mirror dances, "on a purely empirical basis," anyone watching this behavior would suppose that the robots were programmed with mechanisms for detecting objects, pushing them in a given direction toward other objects, and then clustering them together.

In fact, the robots are fitted with only a single kind of sensor (so they are even simpler than Elsie and Elmer) that can detect objects in close proximity, and the robots have only one simple control rule: if the sensor on the right is activated, they turn left, and if the sensor on the left is activated, they turn right. What this means is that the robots have only the ability to avoid obstacles, and yet, whenever they are placed in a cluttered arena, they cluster the cubes together and tidy up. To understand how this works, we have to move beyond the didabot's internal structure and consider its external physical structure and its interaction with the environment (although the tortoise behavior described above should give you a hint).

A didabot has two proximity sensors that are positioned at the front end of its "body," one on each side. The sensors are also placed at an angle, rather than pointing straight ahead. As it moves forward, a didabot can detect cubes off to the side, but, because of the way the sensors are positioned, it can't detect anything directly in front of it. This means that, although a didabot will turn away and avoid cubes on either side, a cube directly in front of the didabot gets pushed along, because the didabot can't "see" it (i.e., its sensors receive no stimulation from it) and so doesn't take any action. If, as it wends its way around the arena, pushing the cube, the didabot's sensor should detect another cube off to the side, it will produce its standard avoidance behavior, moving off to the left or right. But as it does so, it will leave behind the cube it had just been pushing. It will have formed a minicluster. The presence of this larger—and so more detectable—minicluster in the environment increases the chance that another cube, being blindly pushed around by another didabot, will also be deposited in the vicinity. As the cluster grows, so the chances of other cubes' being added continues to grow, until they all are heaped together (with just a stray few here and there around the edges).

It should be clear that this "clustering" is not the "goal" of the individual didabots, nor do the didabots have to know what the others are doing to coordinate their behavior (indeed, they don't know that any other didabots are there at all!). Instead, clustering is an emergent property of a very simple "self-organizing" process that reflects the coupling between the didabots' movements and how these robots interact with the objects in the environment.[18] Most crucially of all, clustering behavior occurs only because the sensors are placed at an angle: move one of the sensors around to the front, and the clustering behavior disappears entirely. This is because objects directly in front of the robot are now avoided in the same way as those on the side, which means no pushing behavior. With no pushing behavior possible, no clustering can occur.

Another extremely relevant lesson we can learn from the didabots is that complex behavior can be produced, not only by a very simple mechanism, but also by a mechanism that bears absolutely no relation to the behavioral outcome produced when that mechanism operates in the real world (after all, who would think obstacle-avoidance was a good way to produce clustering behavior?). Poking about inside a didabot to identify the nature of this mechanism won't tell us anything about didabot behavior because it makes sense only after we have taken into account the interaction of the internal mechanism with the physical structure of the didabot and the structure of the environment.

Does Mate Choice Require Brain Power?

One final robotic example helps flesh out this idea of complex behavior as an emergent property. Barbara Webb's cricket robot[19] is a classic example of the way in which smart behavior in the world is generated by the interaction of neural, bodily, and environmental factors. Unlike Elsie and

Elmer, the first versions of the cricket robot did not physically resemble a real animal; this robot was made of Lego and, like the didabots, looked much like a small toy car. Appearances aside, it was, in every other way, set up to respond to environmental stimuli just as a female cricket would, and, in particular, to produce a response to male mating "songs."

In the previous chapter, we briefly considered how male frogs produce calls that attract mates to them, and male crickets do exactly the same: they produce a "song" made up of a series of chirps. These chirps are, in turn, composed of small bursts of sound that are known as syllables. These syllables have a specific rhythm and are produced at a specific frequency depending on the species of cricket. Studies of real crickets have shown that females orient toward the sound of male songs and approach them (a behavior known as "phonotaxis"). More specifically, females move toward males that have the loudest songs. As only high-quality males can sustain loud chirping, females choosing such males are ensuring that their offspring will inherit the "good genes" that allow the males to outcompete their rivals in this way.

At first sight, female cricket mate choice would seem to involve a series of quite complex mechanisms: a female cricket must first identify a male song against the background of other sounds in the world and analyze the pattern of the song to identify the one that indicates a male of her own species; she then has to compare the males to identify the one that is singing the loudest; finally she has to find and approach the chosen male. How could female crickets achieve all this with the few neurons they have at their disposal? Webb used her cricket to explore the kinds of mechanisms that could underlie this ability, and to show that recognition, comparison, and response need not be three distinct, serial processes but could be achieved by one simple mechanism that capitalized on the physics of the cricket ear.[20]

A female cricket has two ears (or, more precisely, eardrums), one on each of her front legs. These are connected to each other by a tube (trachae). In addition, crickets have auditory spiracles on each side of the top of their body (small openings similar to those used in respiration). These spiracles are also connected to the eardrums, and to each other, by a tracheal tube. At the risk of oversimplifying matters, this means that crickets pick up sound both directly and externally (the eardrum vibrates because of sound waves coming directly from the sound source) and internally, where the sound waves take an indirect route to the eardrum, traveling via the spiracles and trachae.

This arrangement means that the cricket ear is inherently "directional." Let's say that a sound source is emitting sound waves to the left side of the female cricket. The sound arrives at the female's left ear twice, because the sound waves that are picked up externally and internally must travel different distances to reach the ear (it takes longer for the internally detected sound waves to arrive at the eardrum). At the right ear, by contrast, the sound waves arrive at the same time because the two sounds travel the same distance (because the external ear is farther away from the sound). As a result, the external and internal sounds are out of phase with each other on the side closer to the sound source, but they will be in phase with each on the side farther away from the sound source. This, in turn, means that the eardrum on the side closer to the sound vibrates more strongly (has a higher amplitude) .

Each eardrum is connected by around fifty neurons to the rest of the cricket's nervous system, and these converge on a small number of interneurons.[21]. One pair of interneurons in particular is crucial. Each neuron in the pair fires when it reaches a certain threshold, and this is determined by the amplitude of sound. As the interneuron connected to the ear closer to the sound receives a stronger input (because this eardrum is vibrating at a higher amplitude), it reaches its threshold faster and so it fires more quickly than the interneuron connected to the ear that is farther from the sound source.

Neurophysiological studies have shown that crickets always turn to the side on which the neuron is responding more strongly. This means that a female cricket can potentially steer toward a male simply by turning and moving in the direction of the neuron that fires first after each chirp. In other words, instead of analyzing the whole pattern of the syllables and chirps, the female may adopt the much simpler tactic of responding to just the beginning of each chirp. Why, then, does the male song have a distinctive rhythm and temporal pattern? The answer seems to lie in the activation profiles of the females' interneurons. Interneurons show a "decay period" after they fire: they return to their "resting state" only gradually. During this decay period, the neurons are closer to their activation threshold than they are when at rest. This means that if the eardrum is stimulated during the decay period, the neuron will reach its activation

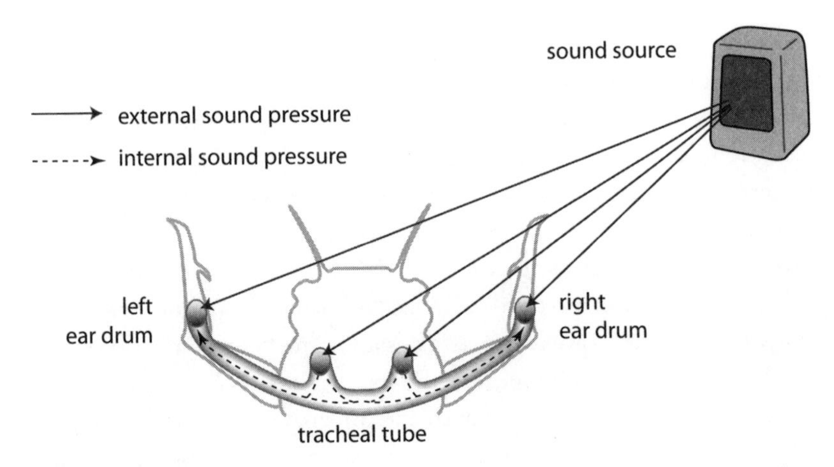

Figure 3.1. Crickets have an eardrum on each of their knees that can pick up sound waves from the environment, and they also have auditory spiracles on their bodies through which sound waves can travel.

threshold more quickly and so it will fire again, even though the amplitude of sound may be lower than that usually needed to trigger a response. So if the syllables in a male song were very closely spaced—and therefore shorter than the decay period—it would start to become much less clear which neuron actually fired first for any given burst of sound, and females would no longer be able to steer accurately toward a particular sound source. Equally, if the gaps between syllables were too long, the female wouldn't be able to track a male accurately because the information would come too slowly: the female would drift off course between the bursts of sound. Male song is therefore "tuned" to the females' interneurons, with the result that females automatically pick out male songs of the correct type against a background of other sounds.

The tracheal tube is also instrumental in all this. Owing to its structure, the sounds that the trachae transmit best are those of the male's calling frequency. Other sound frequencies aren't transmitted so well through the trachae, and so a cricket's auditory system "ignores" sounds of other wavelengths and doesn't produce a directional response. This means that females are not "picking out" and then tracking a male's song against a background of other kinds of songs and noise; rather, they simply don't perceive the background. Instead, they hear and track only the songs that are "relevant" to them. As with Elsie's "mirror dance," the environment—in this case, the male's song—is just as crucial to the feedback process that leads to mate tracking as is anything that happens inside the female cricket. The interaction of the cricket's body (the design of its ears and the transmission properties of the trachae), its "brain" (the tuning of the interneurons), and the environment (the timing of the male song) may allow females to solve all the apparently complex problems of mate finding by following one simple, implicit rule: turn to the side that fires first (it is implicit because there is no sense in which the female knows she is following such a rule, or realizes this is what she's doing). The strength of this "onset" hypothesis is also its weakness, however: could such a simple mechanism really account for all the different kinds of behavior that females display during phonotaxis?

This is where the cricket robot comes in. Webb wired her cricket to mimic the response that a real female cricket would produce if it were using the onset of neural firing to direct its movement. The cricket robot was propelled by two wheels at the rear, each powered by its own motor, with a caster at the front. It was also kitted out with miniature microphones that acted as its "ears," and some electronic circuitry that mimicked the interneurons, one for each side of the cricket. When the circuit for a given ear reached a given threshold, the motor on that side of the cricket was turned off. As the wheels on the other side kept moving, the cricket would therefore turn in the direction of the "neuron" that had "fired" first. In addition, the cricket robot responded only to bursts of discontinuous sound that repeated frequently, to mimic the way that a real cricket's interneurons' recovery times serve to limit a female's response.

To test how her robot responded to male songs, Webb set up a speaker at one end of a small arena (similar to the setup used to test real females crickets) and broadcast sound toward the cricket robot, which had been placed on the opposite side of the arena. When sounds of the optimal frequency and rhythm for the neurons' firing rate were played (the sounds had to be broadcast at a lower frequency than real cricket songs because the robot's "ears" were farther apart, and they also had to be played at a slower pace because of the slow processing speed of the robot's circuits), the cricket moved toward the sound, as predicted by the onset hypothesis. Moreover, the robot followed a zigzag path that looked very like the trajectory shown by real female crickets. When the "wrong" kind of sound was played, on the other hand, the robot would fail: if the rate of syllable production was increased, it simply moved in a straight line and failed to reach the speaker at all, suggesting that the robot couldn't distinguish the gaps in the sounds and so not enough turning signals were produced. If the rate was decreased, then the robot again took fewer turns, but managed to move closer to the speaker in a distinctive curved path, although it often failed to actually reach it. These failures were even better evidence for the onset hypothesis because the same patterns of failure are also shown by real female crickets when they are exposed to sounds that deviate from real male songs in this way.[22]

The robot's ability to recognize and approach a singing male therefore seemed to capture many of the relevant behaviors involved in female mate choice using only a simple mechanism relating to the differential sensitivity of the robot's "ears" to directional sound. But what about the ability to choose one song over another? Webb decided that she would try some mate-choice experiments with her cricket robot, again similar to those performed on real crickets. This time she placed two speakers in the arena and played back a song from each, one louder than the other. Like a real female cricket, the robot approached the "male" with the loudest song. As with the didabots' clustering behavior, the females' "mate-choice" behavior was a truly emergent property; that is, there was no "choice" mechanism explicitly programmed into the robot; it showed this behavior purely as a consequence of the way its internal mechanism interacted with the environment.

Since these first successes, Webb and her coworkers have improved on the cricket robot design, using a more advanced kind of wheeled robot (known as a Khepera robot) that is able to process at faster speeds. This has allowed them to test the robot with real cricket songs, played at the correct speed, and the results are exactly the same: the robot responds to real songs in the same way that real female crickets do, further supporting the idea that the simple robot mechanism is identifying something highly relevant about cricket behavior.[23] Even more recent incarnations of the cricket robot incorporate the cricket circuitry into a Whegs robot:[24] a robot that has "wheel-legs" (hence "whegs"). These robots look rather like insects (with six whegs) and, indeed, are based on the mechanics of walking in cockroaches. With the Whegs robot, the researchers could move beyond the artificial arena and into the real world, where they found that the robot was capable of tracking a male song over the lumps and bumps of natural terrain, bolstering confidence in the laboratory findings.[25]

Perhaps most convincingly of all, Webb has performed some experiments on real crickets which show that female "steering" responses to sound occur well before the central nervous system would be able to process the rhythm of the whole chirp, which, as per the onset hypothesis, suggests they are responding only to the beginning of a chirp. In addition, when the sound pulses were alternated between opposite directions, the crickets would also alternate their steering behavior between left and right. Again, orienting toward individual sound pulses isn't what one would expect if females were analyzing the entire pattern of song and then choosing the best one; instead choice emerges from the kind of simple auditory steering process Webb used in her robot.[26] In other words, these results again reinforce the idea that the "steering" mechanism and the "picking-out" mechanism are one and the same thing. It is the rhythm and temporal pattern of the male song that simultaneously "steers" the female to its source, and "discriminates" the male's chirps from the background.

Complex Behavior from Simple Mechanisms

> Human beings, viewed as behaving systems, are quite simple. The apparent complexity of our behavior over time is largely a reflection of the complexity of the environment in which we find ourselves.
>
> —*Herbert Simon*

All these robot examples demonstrate very powerfully that complex behavior doesn't necessarily require complex internal mechanisms: some things are greater than the sum of their parts. Moreover, it is clear that, as Grey Walter first suggested, the cleverness and behavioral flexibility of Elsie, Elmer, the didabots, and the robot cricket are contingent on the mechanics and wiring of their sensors, rather than the size of their brains. These examples also highlight another theme that we'll develop further in later chapters: once we begin exploring the actual mechanisms that animals use to negotiate their worlds, it becomes very hard to decide where "perception" ends and "cognition" starts. One reason for this is that we may be thinking about these things in the wrong way, and so make false distinctions; we tend to assume that perception is merely the passive reception of information from the world, especially compared to the active manipulation of information that we assume cognition entails. But perceptual processes are much more active than we think, and, as a result, they are highly instrumental in enabling animals to behave flexibly. In the next chapter, we'll consider this idea in more detail by looking at the behavior of another creature that behaves in a manner that, at first glance, seems way above its neuronal pay grade.

CHAPTER 4

The Implausible Nature of *Portia*

I have drunk and seen the spider.
—*Shakespeare,* The Winter's Tale

Jumping spiders, or salticids, to give them their scientific name, are well known for their jumping behavior, as their name suggests. But this isn't what makes them particularly interesting and unusual. What makes them remarkable is their ability to prey on other spider species that are themselves predatory, and the associated complexity of their hunting behavior. One of the best studied of the salticids is the genus *Portia*, species of which occur mainly in the tropical regions of Africa, Asia, and Australasia,[1] where they can be found inhabiting rain forest.

Extensive and detailed studies have revealed that *Portia* species stalk their prey, engage in "deceptive mimicry" of other spiders, take long, complex detours as a way to better position themselves for prey capture, and sneak up on their prey using natural disturbances as smoke screens or even create such diversions for themselves. Moreover, they do all of this with a brain the size of a pinhead. It seems impossible, but as the old joke goes, no one has told the spiders that. Of course, it's not so much the implausible nature of *Portia* that is the problem here, but a failure of our imagination and a hopeless bias toward the big brained. As you will discover, understanding *Portia* spiders means seeing the world from their perspective, quite literally. Before we get to that, however, it is worth expanding a little on how *Portia* spiders go about their business, so that we can get a flavor of the flexibility and contingent nature of their behavior.

Stalking, Sneaking, and Smoke Screens

> Up jumped the cunning spider, and fiercely held her fast.
> He dragged her up his winding stair, into his dismal den.
> —*Mary Howitt*

Observations in the wild have shown that, when a *Portia* spider first enters the web of another spider, it doesn't approach the web-owner directly. This makes obvious sense, because web-building spiders are acutely tuned to detecting the slightest movements of their web. Indeed, most web-building spiders have very poor eyesight, and web vibrations are the primary means by which they "see" and identify their prey. Attracting a web-owner's attention can be fatal because web-building spiders are just as keen predators as the *Portia* species hunting them; *Portia* species need to ensure that they get close enough to the prey to deliver a venomous bite without becoming prey themselves. To do this, a *Portia* spider exploits the sensory machinery of the web-owner by engaging in "aggressive mimicry": it begins to pluck the web with its different legs and palps, varying the speed and amplitude of the signals until it hits on a variant that is sufficiently similar to a real prey item to cause the web-owner to respond. As the web-owner approaches, the *Portia* spider continues to signal, reeling in its prey.[2]

Experiments have shown that a *Portia* spider can produce an almost limitless variety of web signals by varying the number of legs used, plucking with its palps as well as its legs, vibrating its abdomen, and combining the movements made by different appendages in different ways. The spiders can also adjust their response according to the feedback they receive from the web-owner, or, to put it more simply, they engage in a process of trial and error to achieve their results.[3] Initially, when a *Portia* spider encounters a web, it produces web signals randomly until it hits on a variant to which the web-owner responds. The *Portia* spider carries on with this particular signaling pattern as long as the web-owner continues to respond. If the web-owner should stop responding, the *Portia* spider initiates a new sequence until it triggers another response.[4]

As well as mimicking the movements of real prey items, *Portia* spiders can also exploit other kinds of disturbance that cause web movements. If a web is shaken by the wind, a *Portia* spider will begin to step rapidly across the web, taking advantage of the wind-induced smoke screen to cover its own movements. Experiments have shown that *Portia* species catch more prey during these disturbances than in control conditions, supporting the suggestion that this behavior is a predatory tactic.[5] This was further confirmed by an ingenious experiment on the species *Portia fimbriata*, where the spiders were induced by visual cues alone to show smoke-screen behavior. A stationary web—on which the *Portia* spider and a prey spider were placed—was sandwiched between two other webs that were made to shake. The *Portia* spider could see the shaking, and this induced it to produce the smoke-screen behavior, but, as the prey spider's own web was stationary, it could detect the *Portia* spider's approach. Under these conditions, the spider prey showed their classic defense behavior (i.e., leaving the web entirely), and the *Portia* spider was much less successful at hunting.[6] Smoke-screen behavior therefore serves two purposes: it helps *Portia* species stalk their prey more effectively, and it helps to prevent the prey spiders from defensively leaving the web before the *Portia* species is within range for an attack.

Smoke-screen behavior, like aggressive mimicry, is also context dependent. *Portia* species do not use smoke-screen behavior when there are no spider prey present in a web, nor do they do so if they are stalking nonspider prey (which are defenseless and unable to get away). They will, however, use smoke-screen behavior if they detect prey in one of its own webs, and they continue to show the behavior even if they lose sight of their prey item. *Portia* species can also inhibit the "irritation" response that they produce in reaction to particularly strong wind disturbances if prey are present. This irritation response, like smoke-screen behavior, involves more rapid locomotion and more web vibration, but is less guarded than smoke-screen behavior. Their ability to inhibit in the presence of prey shows that smoke-screen behavior is a predatory tactic, and not just a fortuitous by-product of the irritation response.[7]

Portia spiders show similar sensitivity to context when they are hunting other kinds of salticid spiders, and solve the problems they encounter in an equally flexible manner. When hunting these more mobile prey, the *Portia* spiders use the chemical and odor cues produced by spiders' draglines—the lines of silk they trail behind them as they walk. These cues help to prime the *Portia* spider's visual sense so that it becomes more effective at visually locating the particular prey species in question. In one species, *Portia labiata* females can discriminate between their own draglines and those of conspecifics[8]—a form of self-recognition (a facility greeted with great fanfare when it is shown in a mammal besides humans . . .)—and they can also discriminate between familiar and unfamiliar spiders.[9]

When these kinds of chemical and odor cues are not available, *Portia* spiders can also "hunt by speculation":[10] they will suddenly jump into the air and then freeze. These sudden jumping movements are detected by other salticids—all salticids have exceptionally good vision and can pick up the slightest movement: see below—and they then orient in the direction in which they detected jumping. These orienting movements allow hunting *Portia* spiders to detect where the prey is, but because the hunting spider itself is now entirely still, the potential prey cannot detect the hunter. Once the prey has turned away, and is unable to detect any movement, the *Portia* spider begins to stalk.

Compared to when they are hunting nonpredatory spiders, *Portia* spiders move much more slowly and in a more exaggerated fashion when stalking other salticids. The slowness and the exaggerated gait (which helps *Portia* spiders to camouflage themselves by mimicking the way in which detritus on the forest floor flickers in the light) are necessary if they are to avoid detection by these more visually sensitive prey. If a potential prey does orient toward a *Portia* spider at any point during the hunt, the *Portia* spider will freeze as soon as it detects the prey's large, forward-facing eyes (one of several pairs that salticids have: see below). If the prey fails to detect anything and turns away again, the *Portia* spider will begin moving again as soon as the prey's eyes are no longer visible.[11] In this way, it can eventually move close enough to attack and paralyze its prey.[12]

Again, the flexibility of this predatory behavior is impressive, and may be particularly important for a species whose prey items are well equipped to detect and respond to the *Portia* spider's tactics. Hunting other predatory spiders may require a certain amount of flexibility because the *Portia* spiders cannot rely on a single one-size-fits-all response. This is similar to the "social intelligence/social brain" hypothesis put forward to account for the increased brain size of primates: as we've already noted, the pressures coming from the social world, where many social actors are all trying to achieve goals that are only partially compatible with those of others, are argued to have selected for increased brain power to enable animals to plot and plan, and generally to outwit the competition in a highly cognitive "Machiavellian" manner.[13] The spider example, however, suggests that perhaps we could look at this idea slightly differently; selection may not have acted primarily and directly to produce cognitive resources of a particular kind (that is, on an ability to form complex internal representations of the world, and to manipulate them in ways that allow for the planning of future events), but instead it may have acted to produce behavioral flexibility itself, which may or may not require these particular kinds of high-level processes. The spiders demonstrate quite clearly that surprisingly flexible behavior can be achieved in the absence of a large brain.

Taking the Long Way Round

> Adventure is just bad planning.
> —*Roald Amundsen*

While all these hunting tactics are noteworthy, perhaps the most impressive behavior in the *Portia* species' repertoire is their ability to take detour paths through their forest habitat while on the hunt for prey. To take a detour—that is, to move around an obstacle to reach a goal—requires that an animal move out of sight of its goal. This is rightly seen as cognitively challenging because it suggests that a route around the object needs to be planned. The idea of planning is usually taken to imply that an animal must continue to hold an internal representation of the goal during the period in which the goal is out of sight. The complex topography of the forest habitat means that salticid spiders have to make frequent detours before they can get within range of their prey, and, by any standards, the detours of *Portia* spiders are impressive in both length and complexity, rivaling those of many vertebrate species.[14] Here, then, is a real conundrum: how does such a small brain achieve such a remarkable feat of planning?

Early work on detour behavior suggested that salticids used "insight" to work out their routes:[15] they assessed the situation in their heads, made various kinds of plans, and then had a "Eureka!" moment when the answer suddenly became clear. This understanding was probably prompted by the way in which salticids engage in "scanning" behavior prior to making a detour: they turn back and forth slowly, visually inspecting all the routes leading to the prey. During scanning, the spider gives every impression of weighing up the routes for their suitability, planning its way around obstacles, and then setting off once it has worked out a suitable route. But is this really what the spiders are doing? Just because it looks like planning, in ways that make sense to us, doesn't mean that the spiders are necessarily operating in that way. As we noted in chapter 1, the key to understanding other animals besides ourselves is not simply to assume what their capacities are, but to leap in to investigate what is actually the case. For *Portia* spiders, studies of this nature have revealed that the key to understanding the detour and hunting behavior of the salticids does not lie in foresight or planning abilities, but in its perceptual abilities. Specifically, we need to consider, quite literally, how these spiders see the world.

The Eye of the Spider

Salticids have eight camera-type eyes (i.e., eyes similar to ours with a lens to refract light and a retina upon which the image forms, as opposed to the compound eyes of insects).[16] These are evenly spaced around the cephalothorax, which is the front part of the spider's body. There are six so-called secondary eyes, which are arrayed around the sides of the cephalothorax and detect movement, and two principal or anterior medial (AM) eyes, which face forward and can detect fine detail and color. The AM eyes are much larger than the secondary eyes and have a more complex structure, consisting of a large corneal lens embedded in the cuticle of the cephalothorax, behind which is a long tapering eye tube. At the end of the eye tube lies the retina. In other words, a salticid's AM eyes are rather like a pair of binoculars. Toward the end of the eye tube there is also a secondary lens—a concave pit that lies just in front of the retina. The pit functions as a diverging lens because it magnifies the image from the corneal lens, so that, in effect, the eye operates rather like the telephoto lens in a camera.[17]

Unlike human eyes, where the retina forms a single plate, in salticid eyes the retina is arranged into four different layers, stacked one behind the other.[18] As light enters the eye, the light is split into different colors by the corneal lens and the secondary lens. The different colors (i.e., wavelengths of light) are focused at different distances, and these correspond to the different positions of each of the retinal layers. One layer in particular, the "green layer" or layer 1, is important in maintaining a good-quality focused image. In the human eye, we adjust for the fact that objects at different distances in front of the eye come into focus at different distances behind the lens by "accommodating"—we bring things into focus by changing the shape of the lens (and we're performing the same kind of focusing operation when we adjust a pair of binoculars).[19] Salticids cannot focus their eyes in this way, and they solve the problem completely differently. First of all, salticid AM eyes are "active"; they can move their eye tubes around by means of six muscles that enable horizontal (through 60°), vertical (up to 30°), and rotational movement. Second, the different parts of the green layer of the retina are positioned on a kind of "staircase" at different distances from the lens. With this arrangement, the spider can focus on an object by moving its eye tubes from side to side while the corneal lens remains static. As it moves its eye tubes, the spider sweeps its staircase retina across the image generated by the corneal lens. In this way, any object (whether very close or very distant) will be brought into focus on some part of the green layer staircase.

The ability to move their eye tubes may also explain how salticids manage to engage in such complex vision-guided behavior despite such a narrow field of view (about 5° wide and 20° high).[20] In human terms, this is rather like being in a dark room with a powerful flashlight: everything lit up by the narrow beam of the flashlight will be seen perfectly, but almost nothing beyond it. When placed in that situation, one naturally moves the flashlight around, so that the narrow beam scans over the area, and this is what salticids are doing when they move their eye tubes. One suggestion is that the spiders may be able to detect and discriminate particular kinds of objects and features in their environment by moving their eye tubes in complex patterns that are specific to those objects.[21] In this way, the eye itself acts a filter that excludes irrelevant information, a task that would otherwise have to be achieved by neural processing; spiders can compensate for their small brains by having their eyes do most of the work (for more on this, see chapter 9).

Salticid eyes therefore function in much the same way as those of a predatory mammal, like a cat: the secondary eyes, which have a very broad field of view, but low resolution, detect peripheral movements, while the AM eyes detect the fine details of objects directly in front of the spider. The AM eyes are therefore similar to the fovea of the mammalian eye (the region of the retina with highest sensitivity and therefore the highest acuity), while the secondary eyes are functionally equivalent to the mammalian peripheral retina.[22] This gives these tiny spiders their amazing powers of discrimination and detection, despite their possessing only a few thousand photocells (compared to the 150 million humans possess) and a few thousand neurons.

The reason for splitting these functions between two different kinds of eye is thought to be a result of an evolutionary compromise between the need for good vision and the inherent limitations of size that a spider-size body presents. To convert the four salticid eyes on each side of the spider's body into a spherical vertebrate-style eye would require so much additional volume (approximately twenty-seven times more) that a single spherical eye would occupy the whole cephalothorax. The one drawback of having two kinds of eyes is the speed at which perception takes place. While these spiders can detect objects at a distance and make very fine discriminations between them, they can do so only slowly: it takes a long while for a salticid to scan over the cornea with its eye tubes. If good eyesight is taken to be the ability to see in fine detail and color, then salticids can give mammals a run for their money. If good eyesight means seeing things quickly, then one would have to say that salticids see very poorly compared to mammalian predators. As we noted above, it all depends on one's criteria for success.

If we take salticids on their own terms, it's clear that this arrangement—with different eyes playing different roles—serves the spiders well. They can detect movement from almost any angle using their secondary eyes, which makes it difficult to take them by surprise, and they then turn so that the object falls within the field of view of their more complex AM eyes, which pick out fine details. The turning behavior to bring objects into the range of the AM eyes is achieved by a servomotor mechanism that translates the position of stimulation on the retina of the secondary eye into a certain number of steps taken by the legs. Legs on opposite sides of the body move in the opposite direction, so that, as the spider takes these steps, it also turns a specific number of degrees to the left or right.[23]

It is this turning and fixating behavior—or scanning—that gave rise to the idea that *Portia* species plan their routes before setting off on a detour. When we look at this behavior more closely, however, as researchers have done using a variety of ingenious experiments, it is clear that this rather anthropomorphic notion of planning doesn't really get across what the spiders are actually doing.

Finding the Way to Prey

In an investigation of how *Portia* species, in hunting, decide on a route, spiders are given a choice of two routes to reach a prey item. The apparatus is simple, consisting of two horizontal rampways connected to a central pole. A prey item is placed on the pole, and the spider can access it only by walking up one or the other of the rampways. At either end of each rampway, there is another support pole. To reach the prey, the spider needs to climb up a support pole, walk along the rampway, and then climb the "prey pole." To understand how spiders choose their routes, researchers present spiders with either complete routes (i.e., where both horizontal rampways lead to the prey) or incomplete routes, where the apparatus has been fixed so that one rampway has a gap between the support pole and the prey. Choosing this route would mean that the spider wouldn't be able to reach the prey.[24]

At the beginning of each experiment, the spider is placed on a small raised platform from which the entire apparatus is visible. The scanning and fixation patterns of the test spider are then recorded, and the route the spider chooses to reach its prey is noted. In this way, it is possible to work out how scanning is related to the spiders' subsequent behavior.[25]

When looking at complete routes, spiders tend to concentrate their scanning at both the beginning and the end of the scanning routine toward the route that they will eventually take once they start moving. They also head toward the particular part of the detour route that they have fixated on most during the last period of scanning (e.g., the rampway or the support pole). This latter finding provides good support for an earlier suggestion that, when there is no direct route to prey, spiders select and move toward a "secondary objective"[26]—that is, an object that they can reach directly. Once they reach this objective, they then reorient toward the prey, and, if there is still no direct route available, they then select

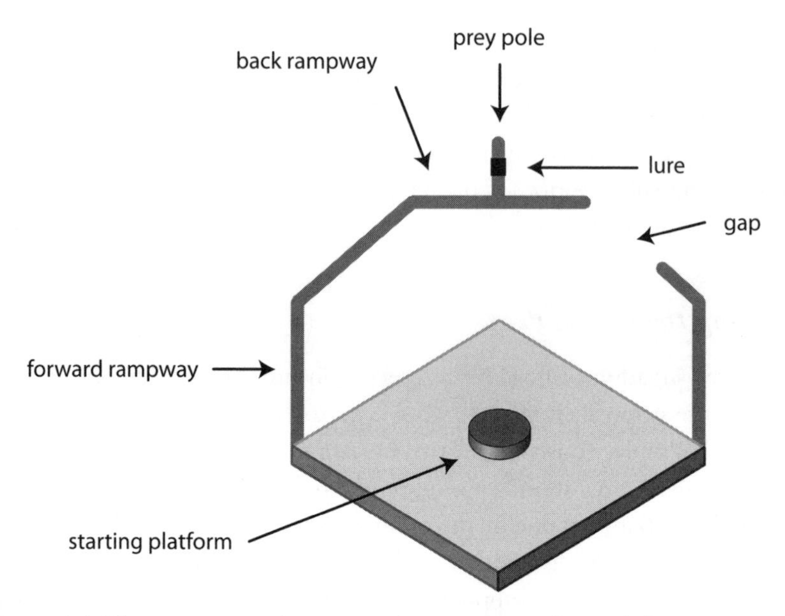

Figure 4.1. The experimental setup used to investigate detour behavior in *Portia* spiders. The prey item is placed on the central portion of the apparatus, with two rampways connected to it, one of which is complete, and one of which contains a gap.

another secondary objective to move toward. In this way, they gradually home in on the prey.[27]

The patterns of scanning and fixation shown in the "gap" condition also reveal a lot about the selection of a detour route. In this condition, the distribution of scanning changes over the course of the scanning period. Initially, the spiders concentrate their fixations on the gap in the apparatus, i.e., on the wrong route. Over the course of the scanning routine, this changes to a pattern of fixation concentrated on the correct rampway that leads to the prey. By the last stage of scanning, the spiders are directing most of their fixations to the complete rampway and— somewhat more strongly—to the pole that connects to the complete rampway, and, again, this is the route they take once they begin moving.

This pattern suggest that the spiders' movements are based on a very simple rule: something like "Move toward the route that was fixated most by scanning" (again, the spider doesn't consciously know it has this rule;

as with the robot crickets and Elsie and Elmer, we can describe its behavior in those terms, but it is highly unlikely that the internal processing of the spider involves an explicit rule of this kind). This in itself isn't very satisfactory as an explanation, however, because it's really just a description of what the spider actually does, not why it does it. More detailed examination of the spiders' behavior, however, gets to the heart of the issue.

Specifically, over the course of scanning, spiders show distinctive variations in their turning direction as they scan the different kinds of routes. The first thing to note is that the spiders tend to "backtrack" away from the prey item as they scan. If they are oriented toward the complete rampway while fixating in this fashion, then the scanning continues in one direction only, with each fixation occurring progressively farther away from the prey. If the spiders are oriented toward the route with the gap, however, they are more likely to reverse their turning direction between fixations, so that they scan back in the direction of the prey, and not away from it.

This pattern suggests that the crucial factor is whether the spider detects an unbroken horizontal line in its field of view.[28] If it detects a horizontal feature, the servomotor mechanism that converts its visual perceptions into movement continues in the direction leading from the prey. If, on the other hand, it encounters the end of a horizontal feature (such as the end of the rampway before the gap), or no horizontal feature at all (i.e., the gap itself), then the spider switches, and, instead of turning away from the prey, it turns back toward it, so bringing the horizontal line it detected previously back into view. Scanning and fixating, therefore, seem to rely on a very simple feedback mechanism that involves two rules: "If the end of a horizontal feature is detected, then change scanning direction," and "If the end of a horizontal feature is not detected, then continue to turn in the direction of the previous turn." These simple rules allow the spider to trace out horizontals leading away from a prey item, and to abandon a particular horizontal route if it turns out to have a gap in it, and so will be unsuccessful.

As with the example of the cricket robot and its mating calls or the didabots and their tidying, the spider doesn't have any understanding or knowledge of the physical properties of gaps: it doesn't understand, nor does it need to, that the presence of a gap won't let it reach the prey, nor does it need to recognize, or know, what a gap actually is, or that such things exist. It just scans and fixates and, as long as a horizontal is detected, it keeps going; if it encounters an edge, it turns and scans back to the prey (which will happen naturally because it will encounter a horizontal all the way). The active movements of the spider's eye tubes over the corneal image may be instrumental in allowing the spider to pick out these horizontal features effectively, and may enable it to "ignore" other features in its environment. As we noted above, the complex pattern of movement produced by the eye tubes acts to filter out most of the world for the spider, allowing it to pick out the most important feature needed for reaching prey. Of course, one can call the patterns of scanning and fixation "planning" if one wishes, but the mechanism used by the spider is very different from the one we envisage ordinarily when using this term, and it is clear that this form of planning has more to do with the size and structure of the spider's eyes than with its brain.

Making It Up as They Go Along

More recently, researchers have looked at what happens with a more realistic setup, where the spider is not placed on top of a platform but is placed in a small hole.[29] Changing the spider's starting position in this way means that the whole apparatus is not visible to the spider at the moment it begins scanning, as in the previous experiments. This has the potential to change how the spider finds its route because the spider can no longer scan the entire apparatus from a single starting position. This setup is more realistic because, under natural conditions, the whole length of a possible detour route is unlikely to be visible to a spider. The platform setup, by contrast, allows the spiders to scan all potential routes to completion and so may create a false impression of how spiders select their detours. Indeed, the platform setup may well have contributed to the notion that the spider "plans ahead" and "thinks before it acts" because the platform ensures that the spiders are in a position to track the entire route to the prey continuously, giving rise to long and detailed scanning routines before the spider makes any kind of movement.

Under these new, more ecologically valid, conditions, there were some very telling differences compared to the results of previous studies. As you might suspect, it was now much more difficult to tell when the scanning period ended and movement began: the spiders' behavior was

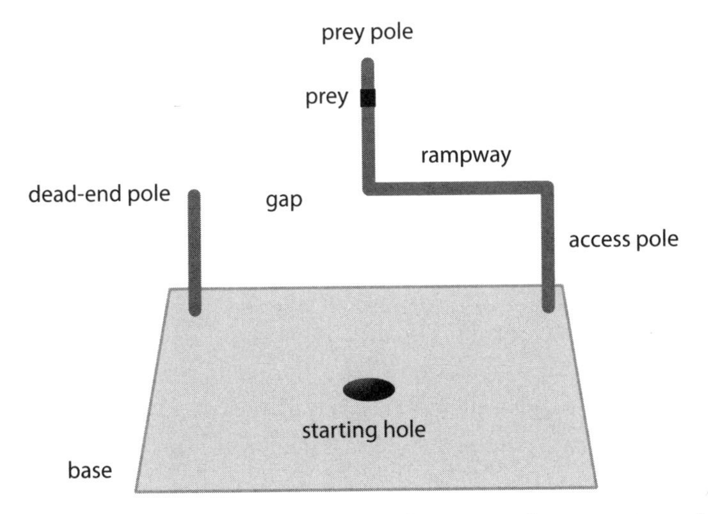

Figure 4.2. The more realistic experimental setup used to investigate detour behavior in *Portia*. The spider is placed in a small hole from which only a portion of the apparatus can be seen.

much more of a mixture of the two: a bit of a scan, some movement, a bit more scanning, and so on. In addition, there was no longer any hint that the spiders had chosen the correct route (i.e., the one with no gap) before they set off toward the apparatus. On the contrary, they were just as likely to head for a "dead-end" support pole (i.e., one that wouldn't lead to prey) as they were to move toward the complete rampway. It was only as their journey progressed that the spiders began to converge on the complete rampway as their destination. Toward the end of their journey, the spiders also developed an even stronger tendency to move toward the support pole leading up to the correct rampway.

This pattern of results suggests that, when spiders cannot scan the whole route in an uninterrupted fashion, they will initially head toward the most conspicuous object they can see, which is why some headed toward the dead end. In addition, it confirms that no planning takes place—if it did, why would they then head for the one item that could not possibly enable them to achieve their goal? Second, as they moved closer to the rampway, the spiders switched from orienting toward isolated objects and began heading toward objects that were connected to each other and therefore likely to lead to the lure. This links back to the previous experiments demonstrating that the spiders traced out the horizontals leading from the prey item: tracing back in this way means that the spiders will automatically encounter any pole that connects to the horizontal rampway. Far from demonstrating that spiders plan a detour route in an insightful way, these detailed and careful experiments reveal that the spiders follow a simple rule that enables them to identify suitable "secondary objectives," and their scanning behavior helps to identify complete and gap-free routes by the simple rules "Keep going if you see a horizontal; turn back if you don't." It is this set of simple mechanisms—not insight—that allows them to reach the prey by a process of successive approximation, and results in the great flexibility seen in their detouring behavior.

The Big Question about Small Brains

> If you cannot do great things, do small things in a great way.
> —*Napoleon Hill*

The flexibility seen in *Portia* spiders, and the emergent and complex behaviors seen in the robot crickets and tortoises, all raise an obvious question: why did brains get bigger in some species? Why aren't all behaviors based on a combination of simple rules that have emergent properties? To answer this, we need to be clear about what exactly brains do for an organism and why. That is, we need to begin placing brains in their ecological context.

Chapter 5

When Do You Need a Big Brain?

I'm a woman of very few words, but lots of action.
—*Mae West*

In *The Day of the Triffids*, John Wyndham's 1951 science fiction novel, giant carnivorous plants roam around England, preying on its human inhabitants, who have been blinded and made vulnerable by a freak meteor shower (I read it at school, and there was also a particularly good BBC adaptation on TV when I was about fourteen, both of which left an indelible mark on me). What makes the story so extremely menacing is the idea of plants that can up sticks and move around of their own accord —plants simply don't do that.[1] One can rely on the fact that a tree, bush, or daffodil will generally remain where you first saw it. Stories like this work because they focus our attention on those aspects of a situation that we usually take for granted—they offer us an alternative that peels back the layers of assumptions we make about the world. Triffids are useful as a way to get us thinking about the links between behavioral flexibility and brains—or, rather, nervous systems as a whole.

We tend to assume that a larger and more complex brain is associated with more complex and flexible behavior, and this has been shown to be true for a number of species.[2] What should now be apparent, however, is that we make such sweeping statements at our peril. After all, the behavior of *Portia* spiders is very flexible—one could almost say inventive—and they don't just respond to the world in a singular, fixed manner. This is true of all animals, from amoebas to armadillos. All show at least some variability in how they act and regulate their behavior in the world. They do so because they are animals, and not plants.[3]

While a dandelion can get all the carbon dioxide and sunlight it needs from one fixed location, the same is not true of an African lion trying to make a living on the savanna. Unlike a plant's energy resources, which are uniformly distributed in space, the large terrestrial herbivores that form the mainstay of a lion's diet are scattered around very unevenly; what's more, these particular kinds of energy resources are prone to head for the horizon the minute they spot a hungry lion on the prowl. To capture these resources, lions must be able to move around and cope with changes in their environment, including the behavior of their prey.

The same goes even for those creatures we tend to perceive as very lowly, like amoebas: to find the resources needed for survival, they also need to detect, contact, and exploit them, and the conditions under which they do this will undoubtedly vary on different occasions. Ditto the fictional triffids; given their need to feed on mobile human prey—even blind ones—the triffids' own mobility is essential. Accordingly they travel widely, and they also communicate with each other and coordinate their actions. Triffids look like plants, but behave like animals, and are all the more creepy as a result.

Even more outlandishly—but this time in a completely factual context—the free-swimming larvae of sea squirts have two life stages: one in which they swim around and possess a small brain (a ganglion, really) of around three hundred cells, and one in which they attach themselves to a rock and then absorb most of their own brain and nervous system, reverting to a more primitive condition—a "process paralleled by some human academics upon obtaining university tenure," as the neuroscientist Rodolfo Llinás once joked.[4] As with triffids, the mobile free-swimming sea squirt needs some flexibility to obtain resources, and so it behaves like an animal, but once it attaches to a rock, it becomes—in effect—a plant. Consequently, if all animals—and triffids—need some degree of flexibility to make their way in the world, then what do we really mean when we speak of more or less flexibility in an animal's behavior? And how exactly do big brains allow them to achieve this?

Instinct and Intelligence

> Let him make use of instinct who cannot make use of reason.
> —*English proverb*

One of the classic distinctions made in discussions of behavioral flexibility is that between "instinct" and "intelligence." Instinctive behaviors are hardwired into a creature's genetic makeup, so that they emerge without any learning or environmental influence (either at birth or at a specific point in development), and they are impervious to change. In other words, instincts are behaviors that occur without thought, are unaffected by learning and memory, and are therefore inflexible. Intelligent behaviors are, of course, exactly the opposite. This makes things very neat and simple, except that, unfortunately, it's not true. Instincts are not genetically specified behaviors that spring forth fully formed, like Athena from the head of Zeus, and they are most certainly modified by learning. Indeed, some "instincts" simply are a form of learning.

Take imprinting, the mechanism by which a young animal follows the first moving object that it sees. Ironically, perhaps, given that it is probably the most famous "instinct" in the world, imprinting involves a form of very rapid learning that occurs just after birth or hatching.[5] The function of this behavior is to allow young animals to form an attachment to their mothers, so that they stay close and gain her full protection from predators and other, less tolerant, members of their own species.

This process must involve learning because evolution cannot select for any kind of inborn mother-recognition mechanism. Individual members of a species differ from each other in a variety of unpredictable ways. This being the case, an inborn recognition mechanism would have to be very general, so that mothers who deviated from the norm would still be recognized by offspring. Any mechanism of sufficient generality to accommodate all the possible variation in female geese would be useless, however: it wouldn't allow the young to discriminate between females, so all female geese would end up being recognized as Mum by a young goose.

A predisposition to prefer stimuli that show a particular kind of configuration—namely, a head-and-neck-shaped region—combined with a period of rapid learning just after hatching, gets around this problem.[6] The innate predisposition to prefer particular kinds of stimuli over others—much like a human infant's perceptual bias for stimuli with top-heavy asymmetry[7]—emerges between fourteen and forty-two hours after hatching, and is itself triggered by experience.[8] This orients the youngster to the right aspects of the environment and provides the basis for learning all the idiosyncrasies of the object and forming an attachment to it. Usually, of course, this "object" is its mother, which is why it works so well. If, however, one interferes in the process, as, famously, Konrad

Lorenz did (work that contributed to his winning the Nobel Prize), then one can produce geese that are imprinted on a human being, and even on inanimate objects.[9] Imprinting experiments demonstrate that our initial impressions of how a behavior is produced can often be completely off target with respect to the actual mechanism, and they also illustrate that the notion of an "instinct" is not as cut-and-dried as we tend to assume in our everyday thinking.[10] Far from enabling individuals to arrive in the world with their knowledge preformed (which is what we usually take instinct to mean) and impervious to learning effects, imprinting is a mechanism that absolutely requires a young animal to learn from experience, both to trigger the predisposition, and to allow the imprinting onto the specific idiosyncratic features of the mother.[11] Indeed, other work has shown that some aspects of this learning begin before the young enter the outside world. Domestic ducks, for example, show a preference for the maternal calls of their own species as soon as they are hatched—something that, again, looks like a hardwired instinct. Experiments have shown, however, that if unhatched ducklings are unable to hear either their own vocalizations or those of other members of their species,[12] then this preference is not displayed after hatching.[13] Once again, learning is a key feature of this "instinctive" preference.

Other behaviors that seem deserving of the term "instinct" because they appear to be innate (that is, present and fully formed at birth) can nevertheless be a product of experience, and can also be altered and modified by any subsequent experiences. Human babies, for example, appear to show a strong preference for the smell of breast milk as little as an hour after birth.[14] Two- to seven-day-old babies also spend longer facing a pad impregnated with the breast odor of their mother than they do an unscented pad,[15] while two-week-old babies, regardless of whether they have been exclusively bottle-fed[16] or breast-fed,[17] orient more strongly toward the breast odor of an unfamiliar lactating woman than toward that same woman's armpit odor or a nonlactating woman's breast odor; all these results suggest that breast odors are generally attractive to newborns, regardless of whether the odor comes from their mothers. While this may look like a fully inbuilt preference (after all, newborn infants have never been exposed to the smell of breast milk), there is some suggestion that it may reflect learning in the womb. Specifically, the smells that the fetus has been exposed to in its mother's amniotic fluid may be similar to those produced in her breast milk, and by the nipple and areola.[18] In turn, the smell of amniotic fluid seems partly to reflect the mother's diet, allowing the baby to learn some unique features of its mother as well as a more general smell "signature": mothers who had been asked to eat anise-flavor during pregnancy, for example, gave birth to infants whose preference for the smell of anise exceeded that of those whose mothers hadn't eaten this flavor.[19]

At their first suckling contact after birth, infants show a distinct preference for a breast that has been treated with amniotic fluid over their mother's natural unwashed breast (and they prefer both over a washed breast). By around six to seven days of age, however, this preference for amniotic-fluid smell has faded, and babies have shifted to a preference for the smell of their mother's breast.[20] If instincts were truly genetically hardwired, both the learning in the womb and the shifts in preference shown after birth would simply be impossible. But these innate responses do show some flexibility, and can be tweaked and adjusted in response to environmental influences.

This brings us back to the example of the *Portia* spiders in the previous chapter: the sensitivity to context and the adaptability they show when hunting their prey make it very difficult to think of their behavior merely as the triggering of a fixed, inflexible response to the presentation of particular environmental stimuli. Even though we have seen that very simple mechanisms potentially can account for their behavior, this doesn't mean that the behavior itself is also simple, or that the creature using these mechanisms should be thought of in a purely invariant, utterly mechanistic way. A lack of brain tissue does not condemn the spiders to behave in a fixed and inflexible manner, a fact that should increase our doubts about any simple distinction between inflexible "instinct" and flexible "intelligence." We need to look elsewhere to understand the kinds of flexibility we're really talking about, and why this particular kind of flexibility evolved.

One promising avenue is to consider more explicitly the context in which animals live and go about their business. As we have seen, the manner in which young animals imprint on their mothers is not wholly internal to the animals themselves; the right type of environmental context is crucially important. The internal physiological mechanisms possessed by the animals are situated in a web of causal influences that have a profound influence on their development and behavioral outcomes. A chick that imprints on its mother and a chick that imprints on Konrad Lorenz may be identical in terms of their genetic inheritance; it is the interaction of this genetic inheritance with their experience of different environmental contexts that leads to the large difference in their behavior. Put more generally, in order for an individual to produce species-typical behavior, it must also inherit an environment similar to that of previous generations, as well as similar genes,[21] and it must undergo a similar developmental process. An abnormal environment can disrupt normal development as effectively as can a mutant gene, if not more so, with major consequences for behavior.

Reliably Recurring Developmental Resources

The thing we need to do, then, is recognize that evolutionary processes depend on the inheritance of a complex of "reliably recurring developmental resources"—that is, all the resources an organism needs to develop the traits that allow it to survive and reproduce—as well as genes.[22] Of course, genes are very important reliably recurring developmental resources, but genetic inheritance alone cannot explain particular behavioral outcomes: the passing of genes from one generation to the next is necessary (you won't get far building any kind of organism without them), but not sufficient to explain the complex patterns of behavior that we see. As we've already noted, one's mother's being the first living thing that one sees is a crucial developmental resource for some species. Sometimes members of another species altogether can be crucial: in mammals, the presence of certain kinds of bacteria is essential for the process of normal gut development.[23] These bacteria are, obviously, not passed on to offspring via genes from their parents, but inheritance by nongenetic means is both evolutionarily stable and reliable.[24] In the case of the woodpecker finches of the Galapagos Islands, particular kinds of physical objects, like sticks and tree holes, are the reliably recurring resources necessary for the development of their tool-using behavior .

The woodpecker finch makes up for the fact that it lacks the long barbed tongue of true woodpeckers by using twigs or cactus spines to pry larvae out of tree holes. (The woodpecker finch belongs to the group known as Darwin's finches, after Charles Darwin, who collected the first specimens during the voyage of the *Beagle*.) The degree to which woodpecker finches display this behavior depends on the habitat in which they live. In more arid parts of the islands, tool use is common, particularly in the dry season, when prey is very likely to be found under bark (presumably to avoid desiccation). In more humid areas, where prey are found in moss and on leaves, tool use is less frequent.[25] A neat experiment on this behavior, using young birds from one of the humid areas, revealed that tool use reflects a predisposition to manipulate and play with twigs during early life, which develops, via learning, into an ability to use tools effectively to gain access to prey (so, again, similar to face recognition and parental imprinting). Humid-area adults exposed to sticks do not show any ability to use tools, however, suggesting that exposure is needed during an early "sensitive period" (parental imprinting again . . .). In addition, social learning isn't necessary for the acquisition of these skills; young birds will learn to use tools regardless of whether they are exposed to an adult "model." What does seem to be important is exposure to sticks and holes containing prey that are inaccessible by any other means. In other words, it is the interaction between the youngsters' predisposition to play with sticks and the availability of holes in which to poke them that leads to effective, functional tool use.

The failure of humid-area adults to use tools most likely occurs because, during the sensitive period for acquiring these skills, they live in a habitat where tool use is unnecessary and also constrained by a lack of tree holes in the environment. Sticks and tree holes can, therefore, be characterized as reliably recurring developmental resources that are essential for fully functional tool-using behavior to emerge.

We can put this in more general terms: instead of considering genes in isolation when we consider evolutionary processes, we can (and perhaps should) think more broadly and consider the entire developmental system as the unit of inheritance: genes, environment, and, most critically, the interactions between them.[26] This doesn't alter the logic of natural selection or undermine the crucial importance of genes for evolutionary change. It just ensures that we keep our eye on the mutual relationship between organism and environment.[27] That is, we commonly speak of organisms' "adapting to the environment," which implies that the environment is "static" and that only organisms undergo any change. But organisms also act on their environments, and are not simply acted on by them, and the environments are therefore changed as a result. Think of earthworms: if they all disappeared tomorrow, the nature of the soil environment would change dramatically. If all trees died, the atmospheric environment would also be altered fundamentally. This is what it means to say that the relationship is mutual: environments act on organisms, and organisms act on their environments. The changes in environments that organisms produce generate new selective pressures as an integral part of this process, and as organisms adapt, they again, simultaneously, change the nature of their environment.[28]

It is also useful to think of organisms and environments as being in mutual relationship because it is becoming increasingly apparent that evolution often exploits so-called self-organizing principles, in much the same way that it exploits the physical closeness between a mother goose and her newly hatched young. There are reliable physical forces operating in the world that can be used by evolution to drive developmental processes in a fashion that does not require any direct or explicit control by genes and gene products. Cell adhesion molecules, for example, are substances that determine the force with which animal cells are held together. Structures of different shapes can be formed through variation of the concentration of cell adhesion molecules along a gradient during development: cells are then pulled together or forced apart according to the force exerted by the cell adhesion molecules. Evolution has tuned the response of the genes to the environment in such a way that the laws of physics—which are very reliable (by definition!)—can determine where a cell ends up; self-organizing processes therefore dispense with the need for any direct or specific genetic control over where a cell should be located. Computer models of these kinds of processes have illustrated how, for example, certain features relating to the evolution of the eye can be explained on the basis of self-organization.[29] Again, this doesn't undermine the importance of genes, but it does help to situate them—to place them in their appropriate context—so that we recognize that genes and environments are in a mutual relationship.

Behavior can also be driven by similar self-organizing principles. Ants are able to discover and follow the shortest possible route to a food source via self-organizing processes.[30] As they leave the nest to forage, ants leave a faint pheromone trail behind them (a substance that attracts others of its species). Other ants show a tendency to follow such trails, especially those with the highest pheromone concentrations. This means that, when an ant finds a food source close to the nest, it returns quickly and the pheromone trail it leaves is strong (because it hasn't had much time to evaporate). A strong trail is more attractive to other foraging ants emerging from the nest, and so, inevitably, these ants also discover the food closest to the nest, laying down their own pheromone trails as they do so. This strengthens the trail even further, which attracts even more ants to follow it and further reinforces the trail. It doesn't take very long for all the ants to converge on the shortest possible route to food, even though there is never any direct communication between the ants, and even though none of them has any idea whatsoever of how food is distributed in the environment. The highly organized, collective behavior is produced purely by the positive feedback generated by successive numbers of ants following and reinforcing the trail. One cannot, therefore, explain this flexible and intelligent behavior by reference to an individual ant, or even to all the ants: it relies utterly on the mutual relationship between the ants and the environment.

The World as We See It

> Our normal waking consciousness, rational consciousness as we call it, is but one special type of consciousness, whilst all about it, parted from it by the filmiest of screens, there lie potential forms of consciousness entirely different.
> —*William James*

What does all this mean for our consideration of flexibility and intelligence? First and foremost, it means that flexibility is a relative concept in two senses. First, when we talk about the behavioral flexibility and intelligence of animals, we're always doing so in relation to the environments in which they are embedded. The flexibility we see emerges as a consequence of the engagement between the organisms and the environment, and is not due to the animal alone (remember the parable of the ant). Moreover, the flexibility and intelligence we see is also relative because, as in our foraging ant example, the intelligence shown is not a property of the animal, or its brain, or even the evolutionary process that gave rise to it, but is, in fact, in the eye of the beholder.[31] We see "intelligence" because we look on as outsiders, whereas, in many cases, the organisms themselves have no knowledge or understanding of the fact that they are exploiting certain physical properties of the environment, or even that such properties exist. Another way of putting this is to say that the degree to which the animal shows flexibility or intelligence is always relative to the frame of reference that is used.[32]

From an ant's perspective, the world consists of varying pheromone concentrations and lumps of food in space, and very little else.[33] It is only from our broader perspective that we can appreciate the intelligence and flexibility of their foraging behavior. This point is extremely useful as it helps us to pinpoint what we really seem to mean when we talk about greater flexibility or intelligence: we deem animals to be more intelligent and flexible when they show evidence of having a greater sense of the world around them, and are therefore able to act on the environment in a larger variety of ways.

This notion of differential perspectives is captured by the concept of the umwelt, a term introduced in the early twentieth century by Jakob von Uexküll, an Estonian biologist. The umwelt is, roughly speaking, the world as it is experienced by a particular organism.[34] Even though a number of creatures may occupy the same environment, they all have a different umwelt because their respective nervous systems are designed by evolution to seek out and respond only to those aspects of the environment that are relevant. For a worker ant, for example, the umwelt is fairly circumscribed because there are only a very few things to which it needs to be attuned in order to achieve success; while the smell of another ant's pheromone trail forms a large part of its umwelt, the smell and appearance of an antelope living in the same environment does not, although it will form a large part of the umwelt of a lion.

Of course, we humans also live inside our own umwelt. We see the world in terms that are relevant to us; we are not sensitive to ultraviolet light, for example, in the way that some insects and birds are, nor are we sensitive to ultrasonic sound, like dogs and bats. These are not part of our umwelt, but because our umwelt is, in other respects, exceptionally broad, we have been able to design artifacts that allow us to detect such factors and understand more about the umwelt of other creatures than

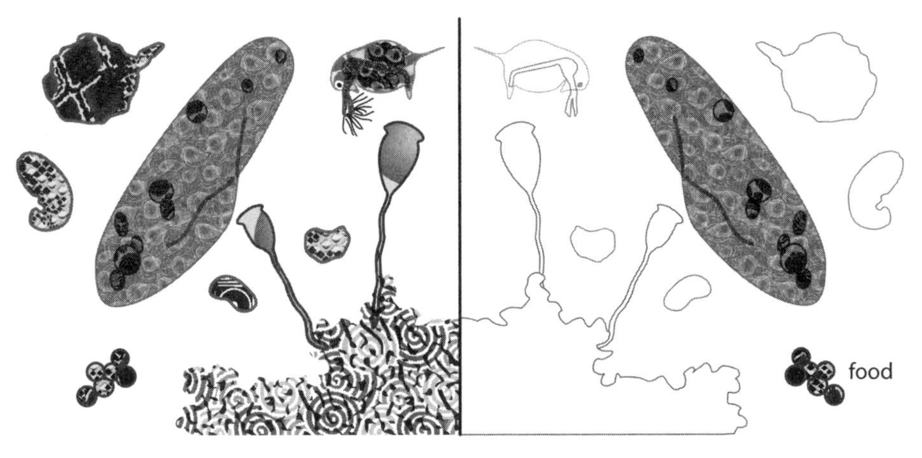

Figure 5.1. The umwelt, or the environment as perceived by the organism. While the environment is full of all kinds of objects and various kinds of stimulus energy, an organism is sensitive only to those that are relevant to its needs.

would otherwise be the case. The umwelt is therefore a valuable concept because it allows us to appreciate both the scope and the limits of species' flexibility, while at the same time preventing us from getting too big for our boots; we, too, have to recognize the limits of our own umwelt.

Taking on board these ideas of frames of reference and the umwelt, we can appreciate that, although *Portia* spiders are indeed flexible with respect to finding their routes to prey, they are, nevertheless, inflexible with respect to what they do with their prey once they find it: they simply eat it, they don't decide whether to bake, boil, or flambé it. Equally, they can't decide not to hunt at all and order in a pizza instead. Different forms of cooking and take-out pizza are simply not part of their umwelt. The flexibility of the spider is, therefore, functionally specific[35] and limited to a particular domain. The same is true of instincts like imprinting: they work exceptionally well in the right domain, but if the environment changes drastically (or another developmental resource is not present or is altered in some way), the imprinting mechanism can go awry because, embedded in their particular unwelt, the animals are unable to perceive that any change has even taken place.

Not So Flexible Friends

A classic experiment by Niko Tinbergen (who shared the Nobel Prize with Lorenz) illustrates this point beautifully.[36] Tinbergen studied how digger wasps find their way back to underground chambers following hunting trips. The wasps dig chambers as a place to lay their eggs, and hunt bees as a source of food for their larvae, bringing them back to their chambers (hence the other name by which these creatures are known: "bee wolves"). Tinbergen was interested in the cues that the wasps used to find their chambers. As with imprinting, the unpredictability of both bee distribution and the most suitable soil for chamber building makes it difficult for evolution to build into a digger wasp a lot of hard-and-fast information about where to find bees or where to build chambers: there is no standard recipe or set of rules that can apply to all wasps at all times. Instead, Tinbergen observed that, as the wasps emerged from their chambers, they wouldn't leave straight away, but would fly in ever-widening circles around the entrance, apparently identifying conspicuous local features that could be used as beacons to guide them back to the chamber: a form of pattern matching that, while simple, is sufficiently flexible to allow the wasp to find its nest.

To confirm his hypothesis, Tinbergen conducted a simple experiment. He placed a ring of pinecones around digger wasp holes; then, once the female had emerged and left on a hunting trip, he simply displaced the ring of pinecones a short distance from the hole. His reasoning was that, if the wasps were using local landmarks, they should fly to the center of the displaced ring, and not to the entrance of their chambers. As predicted, when wasps returned, they would fly to the center of the displaced ring, where they were met with solid ground. As in Lorenz's imprinting experiments, Tinbergen's pinecone trick revealed the limits of the wasps' flexibility: when the world changed too fast, they could not compensate for the change—fast-moving pinecones were not part of their umwelt.

A similar lack of flexibility was also revealed by another of Tinbergen's experiments. This time, when the female went down into the chamber to prepare it, leaving the bee at the entrance, Tinbergen moved the bee a short distance away. When the wasp emerged and failed to locate the bee immediately, it wandered around in the immediate vicinity until it found it and then dragged it to the entrance. Instead of taking it down into the prepared chamber, however, the wasp left it at the entrance, went underground, and prepared the chamber all over again. In other words, the wasp's routine was not flexible, and the wasp couldn't begin in the middle. It did not appear to retain any memory of chamber preparation, nor did it keep track of its own behavior in any way. Rather, each stage of the process seemed to be controlled by specific cues (like the presence of the bee), so that events would always follow each other in a particular order. The completion of one step helped to create the conditions for—and cue the wasp into—the next. Real-time coordination between the wasp's behavior and the objects in its environment ordinarily ensures that the correct sequence of events will always occur; there is, in fact, no need for the wasp to retain any memory of what it has done or what it should do next because, under normal (non-Tinbergian) circumstances, such memory would be a wasted effort: why store internally what is right in front of you anyway? Again, the researcher revealed the wasp's lack of flexibility only by changing the circumstances to a degree far beyond that ever experienced by a wasp.[37]

Another lovely example of how fast-changing circumstances can reveal a lack of flexibility in what seems like highly flexible behavior comes from a study of weaver birds by the zoologist and psychologist John Crook.[38] These African birds build the most wonderful nests of dried grasses, which they weave into a kind of sealed basket—the egg chamber—with a long entrance tube that hangs straight down from the bottom (this is needed to prevent snakes from getting into the nest and taking the eggs). By performing experiments similar to those of Tinbergen, Crook was able to show that the birds that build these delightful objects have no overall concept or sense of their own design.

For example, Crook would slice off the back of the basket-shaped part of the nest while the bird was still building the downward tube on the other side. If the birds were keeping track of the overall design of the nest, they should have simply repaired Crook's damage by reweaving the basket-shaped structure. Instead, the birds would often build a new structure, like another entrance tube, or begin to build a new egg chamber onto the cut edge. This suggests that the birds were not building the nest according to an overall "blueprint" in their heads, but were following simple rules of thumb along the lines of "Where there is an edge, keep weaving until the tube is the right length." What is also important to note from our perspective is that these "rules" were clearly linked to the birds' bodily orientation in space relative to the developing form of the nest: when the exposed surface that Crook cut was at 45° to the original surface, for example, the birds would continue to perch in their original position, making no adjustment for the change in angle, and this would, inevitably, lead to the new entrance tube's being built at 45° to the original one. This further highlights how nest building emerges from the interaction between the birds' bodily activities and the growing structure of the nest itself, rather than reflecting the execution of a preformed plan inside the birds' heads.

Over evolutionary time, of course, any persistent long-term change in environmental circumstances (as opposed to that introduced temporarily by Nobel Prize winners) can result in selection and adaptation to the new circumstances (whether this change is brought about by the organism itself or is some form of climatic change), and the creature's umwelt will also change accordingly. The resulting adaptive fit between organism and environment can be considered as a form of "knowledge"—knowledge that does not sit solely in the animal's head (which is where we assume all knowledge is "kept") but is instead distributed across the whole complex of reliably recurring developmental resources, which together give rise to the behavior.[39] "Storing" knowledge in this way helps to save on expensive neural tissue. Such cost-effectiveness, however, doesn't help a gosling that has formed an abiding attachment to a bearded Austrian ethologist: the bounded flexibility of instincts, which works so beautifully when all the reliably recurring developmental resources are indeed reliable and recurring, falls short when environments change too swiftly or too unpredictably. The inductive logic of natural selection—to generalize into the future what has worked in the past—can work only if both past and future stay more or less the same.[40]

This obviously presents a problem for larger animals, those that live longer, and which reproduce more slowly: they are likely to face a variety of changes in their lifetimes, and relying on the relatively slow process of evolution to select those variants best suited to the new conditions seems a good recipe for extinction. The rate of change may exceed the reproductive rate of the species populations, for example, so that natural selection cannot get any purchase on the problem. Consequently, new variants cannot be generated and selected quickly enough to establish new forms of evolutionary "knowledge."

Luckily, this isn't the only way in which natural selection can act. An adaptive fit between an organism and its environment can also be achieved through selection for a capacity that allows animals to continually update their knowledge of the world. Specifically, natural selection can act to expand the umwelt, so that animals become sensitive to more aspects of the environment and therefore more sensitive to change. It can also increase an animal's capacity to respond to change—and so keep pace with it—by generating a variety of responses to particular environmental contingencies, and then selecting among them for the best solution. The way it does this is to provide animals with a "tracking device"—a brain—that allows them to monitor change, respond appropriately, and learn from the experience.[41] As we've seen, even very simple animals can track certain kinds of changes in their umwelt, but obviously, the more aspects of the environment that need to be tracked so that animals can stay on top of things—and the more perceptual systems this involves, and the more these need to be integrated with each other[42]—the larger and more complex that tracking device needs to be.

The Long and the Short of Behavioral Control

> If everything seems under control, you're just not going fast enough.
>
> —*Mario Andretti*

The main thing a brain does, then—particularly a large one—is give an animal a degree of independence from circumstance: it allows the animal to operate in a manner that, while constrained by selection, cannot be traced in any direct way either to genes or to development. A number of writers on evolution, among them Richard Dawkins, Daniel Dennett, and Henry Plotkin,[43] have all used the example of the Mars Explorer (a spacecraft designed to map the surface of Mars) and the Mars Rover (a robot designed to explore its surface and geology) to illuminate this concept.

When designing the Explorer, and later the Rover, NASA engineers realized that controlling them from Earth would be highly impractical. First, it takes a while to get to Mars—these days it's anywhere from six months to a year—so building a Rover with a finely tuned specific set of instructions is risky. By the time the Rover reached Mars, conditions on the ground could have changed substantially, rendering the current set of instructions useless. Even if all were well, it is still the case that signals from Earth take several minutes to reach Mars, and vice versa: time elapses as signals leave Mars and arrive on Earth, and new instructions are sent back to Mars, with the attendant risk that, by the time they arrive, they are out of date and the response selected is now unnecessary or inappropriate. It should be immediately obvious that this is a situation analogous to Tinbergen's digger wasps, who found that their evolutionary "signals" were suddenly out of date owing to the changes Tinbergen introduced to the environment. With these constraints on an Earth-based control system, NASA's engineers gave up on trying to find a way to keep the robot on this kind of "short leash." Instead, they opted for "long-leash" control, equipping the Rover with a broad set of goals and some fairly general, fairly flexible mechanisms that allowed the robot to control itself, so that it could deal with the particular environmental contingencies it encountered independently of its human controllers back on Earth.

Natural selection acts in the same manner as do NASA's engineers by selecting for long-leash mechanisms whenever environmental circumstances are sufficiently unpredictable that short-leash control is likely to leave an organism in a wasplike "Who moved my cones?" position. A bigger brain gives animals the capacity to acquire, develop, and alter their knowledge according to what is happening in their world at the time it happens. It is important to note, however, that this long-leash control does not simply replace short-leash control wholesale.[44] Rather, these new mechanisms are built on top of the original short-leash ones, augmenting and sometimes changing them: even the brainiest of animals still retain some specific short-leash mechanisms—our examples of face recognition, imprinting, and breast milk preferences all fit this view—largely because evolution follows the maxim "If it ain't broke, don't fix it."

This is an extremely important point, because it is a mistake to think that learning and "intelligence" can operate without "instinct," just as it is a mistake to think that "instincts" operate without any influence from learning (and hence "intelligence"). These extremes simply would not work; animals need both to function effectively and behave adaptively in their respective worlds. We can go further, in fact: it may well be that the greater flexibility of long-leash mechanisms—and the greater complexity of the tracking device that accompanies them—means that animals in possession of such abilities also require a much larger number of innate predispositions than do simpler animals in order to ensure that their environment-derived knowledge does not go off on some useless tangent. It's not a simple trade-off—more flexibility equals less instinct—but a complex interaction between the two. Flexibility is never unbounded.

With these caveats in place, we arrive at the answer to our question: the flexibility we're interested in is that which allows an organism to adopt the principles of natural selection—generating variants, putting them to the test, and selecting the most successful—and use them for its own ends within its own lifetime. As the future becomes more uncertain, so the past becomes a less reliable guide to the future, and an animal must become more responsive to circumstance in order to control its dealings with the world, survive, and reproduce. As the benefits of greater control mount up, so the costs of a larger, more complex brain are balanced out.[45]

An (Old Female) Elephant Never Forgets . . .

To put this in concrete terms, think of the variation in the environmental conditions encountered by a mayfly that lives for just one day, and an elephant that can live for seventy years. On the one hand, a mayfly doesn't have—and doesn't need—a very broad umwelt to get its job of emerging, breeding, and dying over and done with. Elephants, on the other hand, encounter regular seasonal changes in their habitats (which can be more or less predictable and uncertain) along with longer-term shifts that can occur over several years, such as alterations in the water table and unpredictable events such as drought. On top of this, they must contend with changes in their social environment as other elephants are born, mature, produce more elephants, and die. The latter problem is particularly acute because of the nature of elephant society.

Elephant females live in family groups, organized around a single, older female, known as the matriarch. These family groups join up with other groups from time to time, but these larger groupings do not remain stable for very long, and tend to break up and rejoin as families go their separate ways. In Amboseli, Kenya, where the elephant population has been studied for over 30 years, a single elephant family encounters approximately 25 other families a year on average, which amounts to around 175 other female elephants.[46] Some of the these families associate with each other more closely than others, and less familiar groups can often be hostile, harassing young elephant calves. Given the unpredictability with which other families are encountered, and the large number of individuals involved, an elephant's social world is one of continual change, where the action taken depends on a number of shifting variables, like whether a family is alone or has joined up with another, whether the group it detects is familiar or unfamiliar, and whether there are young elephants present, and how many.

The ability to recognize and respond appropriately to different groups has been shown to have important reproductive benefits, and, interestingly, families with older matriarchs are much better at detecting and responding to other groups. Using an experimental playback technique, one can simulate encounters between families by broadcasting recordings of one group's vocalizations (the infrasonic rumbles that elephants produce to keep in contact with each other) to another, making it appear as though the first group were approaching from a distance; one can thereby investigate what makes an older matriarch's group more effective.[47]

The reason why this technique can tell us something about how elephants discriminate between different families is that the females tend to bunch together defensively when they hear the sound of unfamiliar elephants. Overall, families with older matriarchs were much less likely to bunch than families with younger matriarchs, suggesting that older females were more familiar with the calls of more females (owing to their greater experience) and/or they were more socially confident. On those occasions when an old matriarch's family did bunch together in response to unfamiliar females, they all tended to show a much stronger response than did families headed by younger matriarchs: older females thus appeared to be more accurate at discriminating familiar from unfamiliar families, and so were able to protect themselves more effectively from strangers. In line with this notion, the age of the matriarch was shown to be a good predictor of the number of calves produced by females each year, suggesting that these increased discriminatory abilities of older females translated into reproductive benefits for all.[48] The ability both to update their knowledge on a daily basis, and to learn from experience during their lifetime, is therefore key to female elephants' successful survival and reproduction; evolutionary short-leash "knowledge," like that of the digger wasp, simply won't do.

Cache on Delivery

Another animal that illustrates how evolutionary and real-time knowledge can act in concert to produce adaptive behavior is the Western scrub jay. Native to the western United States, these birds are members of the crow family (Corvidae). As they don't migrate for the winter, scrub jays store food (e.g., acorns and other seeds) and use these caches to see them through the cold months when less food is naturally available. Given that winter is a reliable occurrence in the temperate zones of the world, the knowledge that jays use to cache food is the kind of evolved knowledge that can be built into the complex of developmental resources. In addition to this predictable seasonal change, however, the scrub jays have to contend with two other factors that require their knowledge of the world to be updated more rapidly: first, different kinds of food tend to perish at different rates, and, second, other scrub jays may find and eat their cache before they do.

In a series of complex but fiendishly clever experiments, scrub jays have been shown to remember what kind of food they have stashed where, and, more impressively, when they did so.[49] When given the option of recovering either previously cached wax worms (their favorite food—"crack for corvids"—but one that perishes rapidly) or peanuts (which are less preferred, but last longer), scrub jays would attempt to recover the wax worms if the delay between caching and recovery was short (4 hours) but would recover peanuts when the delay was long (120 hours).[50] Wax worms are still tasty after 4 hours, so it makes sense to go for them after a short delay, but after 120 hours, they have gone bad and become inedible, and there's no point in recovering them. It seems, then, that the scrub jays have some means of keeping track of what they have cached, where they did so, and also when they did so, with the result that they do not waste effort recovering bad food, or, conversely, they don't spend time and effort recovering less preferred nonperishable foods, like peanuts, when there are more exciting and tasty foods like wax worms to be had.[51]

In addition to recovering their own caches, scrub jays are also very adept at pilfering the stores of other scrub jays. To counter such pilfering, jays frequently recache their food if there are competitor birds around that have had the opportunity to observe where food has been stored. In another series of experiments, scrub jays were able to keep track of which food caches had been seen by competitor birds, and even to remember exactly which specific individual had seen them hide their cache. In one experiment, for example, the birds were given access to food and allowed to cache in private or in the presence of an observer bird. After a three-hour delay, the birds were then allowed to recover their caches, and to recache the food in a new tray if they chose to do so. Jays were much more likely to recover their caches and hide them in the new tray if they had been observed initially than if they had cached in private, suggesting they were sensitive to the social context in which they had originally cached: food was moved to a new spot only if there was a risk that it could be pilfered by another bird.[52]

Scrub jays also engage in various kinds of cache-protection strategies at the time of caching: if they can be seen by another bird as they cache, scrub jays are much more likely to choose caching sites in shady areas, or those that are less well lit (presumably because this makes it harder for a competitor to see exactly where they are caching), and they will also choose a more distant site over a closer one.[53] Further, they are more likely to move a food item around multiple times between different cache sites before finally settling on one, making it difficult for a competitor to keep track of exactly where a food item ends up. Most intriguingly of all, it appears that these recaching tactics are dependent on a scrub jay's previous experience of pilfering: the only birds to recache their food after being observed are those that have previously taken food from other birds' caches. Birds with no experience of raiding another bird's cache sites show no tendency to recache their own food if another bird is witness to their caching.[54]

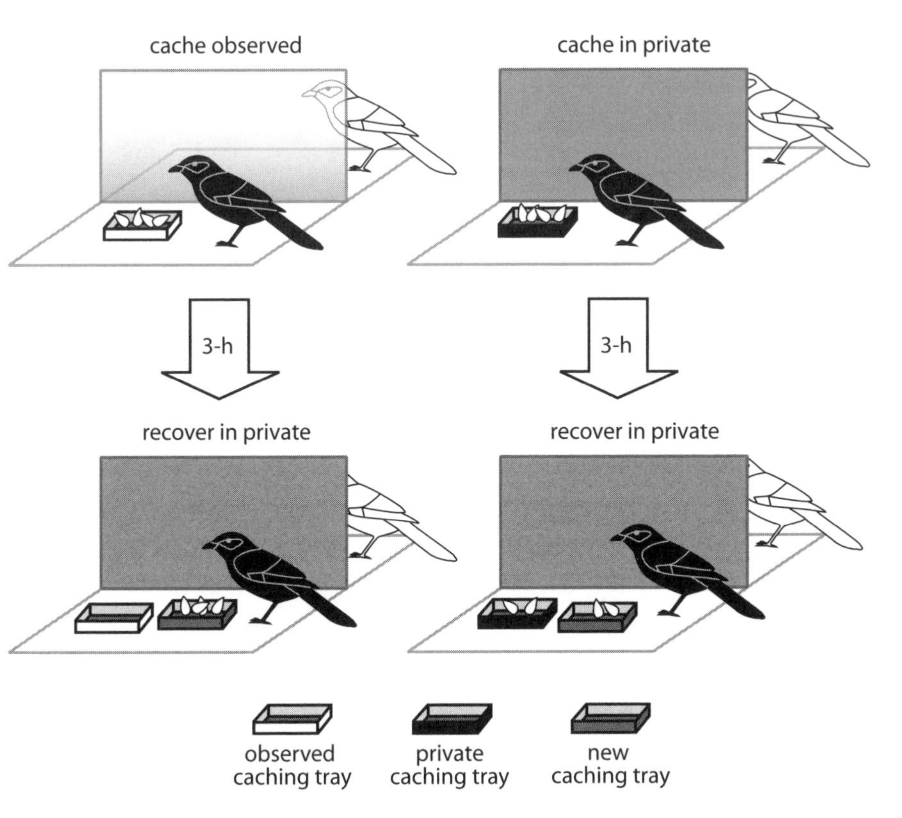

Figure 5.2. One of the experimental setups used to investigate the flexible caching behavior of scrub jays. In this setup, the jays are given the opportunity to cache food in the presence of another bird or to do so privately, and later they are given the opportunity to recache their food.

Scrub jays, then, not only have a finely tuned sensitivity to what was stashed where and when, and whether other birds were present at the time, but it also appears that this ability is honed by their own experience. So, while caching is a behavior under short-leash (evolutionary) control (after all, the birds continue to cache even in a laboratory setting, where they are never short of food and where winter never comes), the more unpredictable elements of caching behavior—differing rates of food decay and the presence of observers—appear to have selected for long-leash, real-time control that enables the birds to regulate and monitor their caches according to their specific local circumstances.

Change from Within

One final source of variability we should consider, briefly, is the way in which many animals have to contend with changes in themselves over time. Let's consider elephants again: a behavior that works well for a very small juvenile elephant may not be so efficient for a very large adult one. Changes in body size, power:weight ratios, and even the digestive system (as an animal switches from its mother's milk to solid food, for instance), all require adjustments in behavior that cannot always be dealt with by short-leash evolutionary control. Many animals, particularly the larger mammals and birds, therefore require greater behavioral flexibility because of changes in the way that their own bodies interact with the environment over time (and, as we noted earlier, this may also require more innate predispositions to help ensure that animals are tuned appropriately to relevant environmental features).

From our perspective here, this last point is perhaps the most important because it brings together two issues that were hinted at in previous chapters. The first is that an animal's brain is part of its body, and that bodies and brains work together to produce effective behavior; it is imperative that we don't lose sight of this fact and think that we can somehow consider brains in isolation from bodies. To be more precise, we should abandon talk of brains altogether and talk about the increasing size and complexity of the nervous system as a whole. It is a mistake to think of the brain as somehow being in charge of the body, with the rest of the nervous system reduced to a set of "message cables"[55] that merely ferry information to and from the brain, and the body reduced to the means by which the brain is ferried around the world. As we'll see in chapter 9, putting a body together in a particular kind of way often eliminates the need for any kind of neural control of behavior; the body itself takes care of things and functions in a very cost-effective way as a result. It seems that, again, our focus on our own large brains has caused us to overlook the possibility that cognition is a property of whole organisms and not of brains alone. Second, we can see that an animal's brain and body cannot be divorced from the environment in which it lives. It is a mistake to think that we can study an animal's behavior and cognition in isolation from the environment. In the same way that a "mutualistic" view pays dividends when we endeavor to understand evolutionary processes, a similarly mutualistic view of behavior and psychology is equally valuable.

CHAPTER 6

The Ecology of Psychology

The idea of environment is a necessity to the idea of organism, and with the conception of environment comes the impossibility of considering psychical life as an individual isolated thing developing in a vacuum.
—*John Dewey*

When was the last time you danced? Maybe it was at a party or a nightclub or a wedding. Maybe you simply danced around your kitchen while doing the dishes, or perhaps you dance for a living, in front of thousands of people. Maybe you hated every minute; maybe you had the time of your life. Now, as you think of yourself dancing, think also about where dance is located. A weird question, right? What does it even mean? Well, let's get more specific. Is the dance inside you? Is it some kind of state that you're in? Or is the dance something simply that happens to you?[1] Makes even less sense, now, doesn't it? Because dances just aren't like that. And that is precisely the point: a dance is a something we do, not a thing we possess, or a state we occupy.

When we dance, we're coordinating our movements with our environment, in the moment; the dance doesn't sit inside us, and it isn't dependent on us alone, but on the music as well, its rhythm and tempo, and also on our partner, if we have one. It simply makes no sense at all to ask where a dance "is." The philosopher Alva Noë uses this dancing metaphor to explain his view of human consciousness; like a dance, consciousness is something we do, not an object we possess. As he suggests, "our ability to dance depends on all sorts of things going on inside of us, but that we are dancing is fundamentally an attunement to the world around us."[2]

We can extend Noë's dancing metaphor to include all psychological phenomena as a whole—not just conscious experience—and we can extend it to all other animals as well.[3] We can, in other words, adopt what is known as the "ecological" approach to psychology.[4] According to this theory, developed by the late James Gibson, psychological phenomena are not things that happen "inside" animals, but are found in the relations between animals and their environments; hence they are "ecological."[5] The idea of psychology as ecological phenomena, along with Noë's dancing metaphor, pins down precisely the mutuality of organism and environment we identified in the previous chapter. In this chapter, we're going to explore the benefits of an ecological approach to psychology, and so the argument from here on gets a little more technical. The ideas we're going to consider are important and interesting, but they do require a bit of effort to understand, not least because they completely turn many of our everyday assumptions on their head, and their unfamiliarity alone may make them difficult to grasp. It is, however, well worth the effort, so bear with me.

Affordances and the "Loopiness" of Behavior

> The agent does not merely receive input passively and then process it. Rather the agent immediately sees things from some perspective and sees them affording a certain action.
> —*Maurice Merleau-Ponty*

> The verb to afford is found in the dictionary, but the noun affordance is not. I have made it up. I mean by it something that refers to both the environment and the animal in a way that no existing term does. It implies the complementarity of the animal and the environment.
> —*James Gibson*

As we've seen, even fairly simple animals explore and regulate their encounters with the environment in highly active ways, exploiting the structure of their bodies and the habitat in order to make their tasks simpler, more effective, or both. The idea of an active organism is key to understanding many of Gibson's arguments because it completely undercuts many of the assumptions we hold about sensation, perception, and action.

For example, one of the first questions Gibson asks is "What are the senses?"[6] We usually assume that our senses are "channels of sensation,"[7] but Gibson's view is that our senses are systems for perception. This perhaps seems odd at first, because the conventional view is that sensations are the "raw materials" from which we then form our perceptions. But, as Gibson points out, the verb "to sense" has two meanings. The first is the one we've just described, "to have a sensation," but the second is "to detect something," and it is this second sense that Gibson uses: for him, perception is based not on sensation, but on detecting information. What is striking, then, about Gibson's approach is that it separates the input to the nervous system that leads us to experience sensations from that which leads to perception. Gibson gives the example of the "obstacle sense" of blind people, who can sense objects as a kind of "facial vision." In fact, such people are detecting objects using echolocation, and so it is their auditory system that they are using. Blind people therefore have a particular kind of perceptual experience without realizing which of their senses has been stimulated. As Gibson describes it, this perception is "sensationless," meaning that the sensations experienced do not actually reflect the mechanism by which the information was detected.

Gibson also questioned the classic psychological categories of "stimulus" and "response." In the laboratory, these categories are appropriate because we can "impose" discrete and individual stimuli on the sense organs in isolation (a pure tone, a flash of light), and the specific response to such stimuli is noted: how intense must the stimulus be to produce a response? How long must it last? How intense is the response? Outside the laboratory, however, individual, independent stimuli of this kind don't exist: they overlap in space and time; they merge together; they change position. What we have in the real world is a "flowing array of stimulus energy"[8] from which animals can obtain information for perception: animals act, move around, change their position, and so alter the nature of the stimulus information available to them as a consequence. In the natural environment, unlike the laboratory, animals are not limited to passively receiving whatever happens to come their way; they can actively seek out the information they need. In other words, their "responses" may often precede any given "stimulus" (although, of course, calling it a response is completely wrong because it's not a "response" to anything). Perception, then, is a matter of active exploration of, and attention to, the environment, and as such, it calls into question the existence of two independent categories of stimulus and response. Instead, it seems more accurate to talk about an ongoing process of "sensorimotor coordination," as the philosopher, John Dewey, called it, with behavior viewed as a continuous, integrated loop of action and perception.[9]

The idea that perception involves the active detection of information also means that we need to think about the senses rather differently. We are used to thinking of the senses as passive receptors for various kinds of stimulation—light energy in the case of our eyes, sound waves in hearing—and so we study "the eye" or "the ear" accordingly, working out the various mechanisms by which, say, sound waves are detected by the hair cells in the cochlea, or the rods and cones of the eye are stimulated by photons. There is nothing wrong with all this—it is fascinating to discover how our receptors work—but, as Gibson noted, the perceptual apparatus by which animals pick up information in the world around them is not achieved by these receptors in isolation. That is, the perceptual system we use to detect sounds in the world is not "the ear" but both of our ears, positioned on either side of a mobile head, attached to a mobile body, connected to our entire nervous system. To detect the source and identity of a sound, we have to move our heads and often our bodies because—as we saw with the crickets in chapter 3—localizing sound depends on the relation between the sound waves that arrive at each ear (sound takes longer to reach the ear that is farther away from the source). An ear can't do this alone; only the whole perceptual system of the active organism can do so. A "perceptual system," then, is not simply a receptor attached to a nerve; it involves the entire nervous system because it requires the whole body to pick up information, not just the sense receptors. Gibson suggested that we should think metaphorically about the senses acting like tentacles or feelers[10] that seek out information through exploration as a means to help get us away from the idea of the senses simply as passive receivers.

Once we think of animals as explorers of their environments—as active seekers of information, rather than simply passive receivers—it raises the question of what it is they seek. Gibson's answer is that organisms seek out information and regulate their behavior with respect to the "affordances" of the environment: the opportunities and possibilities for action that particular objects and resources offer to an animal.[11]

So what exactly are affordances? For a human, of a certain size, with two legs that bend in the middle and a squashy bottom, a chair affords the possibility of sitting, as does a tree stump, but such objects do not afford sitting to a giraffe or a cow; similarly, a fork offers the possibility of feeding oneself for a human, but not for a fish, a dog, or a crow, all of which lack hands. A fig tree affords climbing for a chimpanzee, whereas for an elephant it affords scratching its bottom or pushing over. Perception is, then, "written in the language of action"[12] so that we see not chairs but places to sit; the spider sees not a vertical pole but a place to climb; the woodpecker finch sees not a stick but something to poke with, and it sees not a hole but a place to poke.

The concept of affordance means that what goes on in an animal's head (whatever that might turn out to be) cannot be separated from how it moves its body about in the world.[13] The "loopy" cyclical nature of exploratory behavior forces the realization that perception and motor action do not form two discrete categories, but instead they work together as a single tightly coordinated, fully integrated unit to detect and exploit affordances, and so produce highly specific adaptive behavior. This in turn means that the same environmental resources will offer different possibilities (different affordances) to different organisms, because they possess different kinds of bodies that differ in their sensorimotor capacities.

This ties the notion of affordances to that of the umwelt very nicely. Affordances are "organism-dependent," like the umwelt, because they reflect the degree to which an animal with a particular kind of nervous system can detect and make use of particular kinds of environmental opportunities. This doesn't mean that affordances are purely "subjective," however. As Gibson puts it:

> an affordance is neither an objective property nor a subjective property; or it is both if you like. An affordance cuts across the dichotomy of subjective-objective and helps us to understand its inadequacy. It is equally a fact of the environment and a fact of behavior. It is both physical and psychical, yet neither. An affordance points both ways, to the environment and to the observer.[14]

So a rigid horizontal surface affords walking for animals with legs, regardless of whether there are any animals actually present to walk on it, but, at the same time, the affordance is realized only when an animal with legs exploits that structure in that way.[15] Another nice example is the "Thank God" hold,[16] a term used in rock climbing to describe an attached object on a cliff that affords a safe, secure, and easy grasp. A climber who encounters one of these at the end of a long climb, or during a particularly strenuous stretch, is likely to be enormously pleased, hence its name. Nevertheless, even though there is a relationship between the nature of the hold and the feelings of the climber, the aspects of the hold that specify its "Thank God" qualities are present whether or not anyone is there to use it, and the hold is always there to be perceived and used.

Taking Control

The concept of affordances reinforces the point made in the previous chapter that animals act on their environments, and are not merely acted on by them. The *Portia* spider actively mimics the movements of other spider prey with its web-twanging antics because the web affords such a possibility. Arguing in this way doesn't mean that spiders have any knowledge of why they are bothering to twang another spider's web, or any understanding of why web twanging results in a meal, but it does mean that, as we noted above, we can't make any hard-and-fast distinctions between "instinctive" and "intelligent" behavior. It also reinforces the idea that behavior is not the result of a one-way link that goes from stimulus to response, but a loopy process of "sensorimotor coupling" in which action and movement often precede sensory stimulation. This reversal of our usual way of thinking allows us to recognize that, ultimately, behavior is about controlling one's perceptions.[17]

Consider driving a car within an area with a strict speed limit of 60 mph. To keep your perception of this intended speed constant (whether by looking at the speedometer, or by ensuring that objects to the side move past at a constant rate), you have to continually adjust your behavior by varying the pressure of your foot on the accelerator and brake as you encounter differences in the road surface and gradient. Your action in response to the sight of the speedometer needle climbing will be to increase pressure on the brake, and the reason you do this is that you want your next perception to be the speedometer falling again. This should sound familiar, as it is the same kind of negative feedback mechanism that Grey Walter used in his tortoise robots; in Elsie's and Elmer's case, they were attempting to control their perception of light—not too bright and not too dim—and so their actions in the environment constituted an attempt to produce the right level of sensory stimulus, not the right response. Similarly, your actions with respect to speed are also an attempt to produce the right stimulus, and not the right response, because placing your foot on the brake is "right" only in the context of controlling how the speedometer needle looks to you. When you are going uphill, placing your foot on the accelerator will become the response that produces the right stimulus.

It is much easier to work out what the "right" stimulus will be for an animal (generally that which promotes survival, e.g., the perception of a full stomach, not an empty one; the perception of safety, not fear) in a given situation than it is to decide on the "right" response, because this could vary from moment to moment.[18] Perceptual control theory (PCT)—as this field of research is called—therefore argues that the reason why behavior varies is that animals are trying to maintain stability in their perceptions of the world. So, like Gibson's and Dewey's theories, PCT is also a theory of behavior that considers animals to be "purposeful": an organism controls its own behavior, and hence its own fate, by its actions in the world. Its "purpose" is to defend its internal states (i.e., to sustain homeostasis) and the external state of the perceived world, so that it remains within certain limits that are conducive to its survival.

From the perspective we are developing here, PCT is also an attractive theory of behavior because, like Gibson's theory of ecological perception, it links well to the idea of the umwelt. A controlling organism can know only its own sensory signals or perceptions—it can't look back at itself and know the world outside of its own perception of it.[19] As with the umwelt, then, PCT forces us to remain aware that our observations of an animal's behavior from the outside will necessarily be very different from the behavior as seen from the inside, and we shouldn't assume that, just because we can see and assess certain aspects of an animal's actions, these actions are relevant to the animal itself in terms of achieving its goals.

Take a gymnast, for example. People assessing a gymnast's performance can attend to the outward appearance of her actions, but the gymnast herself is not directly aware of how she appears to others; she controls only her own perceptions (of pressure, effort, sound, sight) that the judges cannot experience or assess.[20] From a classic stimulus-response perspective, however, it is very tempting to explain aspects of the gymnast's outward appearance and behavior that do not exist in the perceptual world of the gymnast herself, so we are then, in effect, explaining what the gymnast is not doing, rather than what she is.[21]

The Environment as Illusion

> [T]he real problem is how the cortex uses the messages it gets from the retina to answer its questions and to ask others. This is the serial process we call visual perception.
> —*J. Z. Young*

> And if the brain, why not the kidney?
> —*Peter Hacker*

Another aspect of Gibson's theory that links to his rejection of perception and action as two separate systems is the rejection of the conventional idea that all the sensory inputs we receive from the world need to be processed, via internal representations, in order to produce a rich and detailed view of our environment.[22] To understand how and why Gibson's theory differs, we first need to understand a little about the standard view of perception as a process. We'll take visual perception as our example.

The conventional view of visual perception originates with the mathematician and astronomer Johannes Kepler, who worked out the optics of the eye (based on the assumption that the eye functioned as a camera does, or in Kepler's day, a camera obscura), and demonstrated that, as a consequence of the way that light rays were refracted by the lens, the image formed on the retina would be both upside-down and reversed.[23] This, of course, raises an interesting question: if perception of the environment begins with the formation of an image on the retina, and the retinal image is inverted, static, and two-dimensional, how can we perceive our environment as three-dimensional, upright, and dynamic? Not only this, but we have two eyes, so there are two retinal images, and these aren't identical, so why don't we see double? Clearly the two upside-down, static, nonidentical images must get converted into a three-dimensional view of the environment, but how? This problem taxed Kepler, but his solutions to it were, it has been suggested "futile and ad hoc"; they made little sense and were highly unsatisfactory.[24]

The French philosopher René Descartes (more of him anon) later conducted experiments with dissected bulls' eyes, and revealed the upside-down image on the retina just as Kepler had predicted. Descartes, however, came up with a solution to the problem of the retinal image that seems entirely reasonable. His argument was that the stimulation provided by the retinal image, and the images then formed on the pineal gland (the part of the brain where Descartes assumed the images from the retina were sent), were simply that: patterns of stimulation that could be processed in various ways to produce our visual experience. Descartes, in other words, played down the problem of the upside-down retinal image by arguing that the stimulation on the retina received from an object in the world didn't actually have to resemble that object, just as the two-dimensional picture or painting of an object in the world doesn't completely resemble the real three-dimensional object it depicts. All that matters in both cases is that the same mental activity is aroused by both.

In the 1800s Hermann von Helmholtz, who is often described (quite rightly) as the Newton of psychology, extended this by arguing that visual perception was a matter of "unconscious inference" by the brain.[25] According to Helmholtz, nerve impulses sent from the retina were transformed into sensations in the brain[26] and acted as the "raw material" from which our perceptions could then be "inferred" by the unconscious mind. Helmholtz reasoned that perception had to be a process of inference because the information sent from the retina was so scanty; it couldn't possibly provide an accurate representation of what the world looks like, and so our brains had to fill in the missing parts. Following Helmholtz, the distinguished neurophysiologist-psychologist Richard Gregory also suggested that our perceptions were "hypotheses" that our brains form about the world, based on the impoverished data received from incoming neural signals.[27] The equally distinguished neuroscientist Colin Blakemore similarly argued that "neurons present arguments to the brain . . . arguments on which the brain constructs its hypothesis of perception."[28] Most famously of all, David Marr, in his seminal book, *Vision*, stated quite emphatically that vision was a process of information analysis conducted by the brain.[29] In essence, then, and as

Noë puts it, the conventional view is that, because it seems that we are given much less than we think we are (the flat retinal image), what we see is not the world itself, but the world we create inside us, using our brains.[30] In other words, we contact the world only indirectly, because we have to construct a detailed, faithful representation inside our heads, and we then act on the basis of this reconstruction, rather than acting directly on the world itself.[31] The environment as we see it is, therefore, an illusion.[32]

As the neurophysiologist Max Bennett and the philosopher Peter Hacker point out, there are several conceptual problems with this approach to perception.[33] Perhaps the most prominent is that it commits the "mereological fallacy": put crudely, this means to treat parts as though they were wholes. To say that the brain "infers" and "hypothesizes" or that neurons "present arguments" is to treat the brain as though it were a person in its own right—one that sits inside your head and then tells "you" things[34]—instead of being only a part of you.[35] Another way to put this is to say that we anthropomorphize our brains. This is inappropriate because the brain is an organ with cells that generate action potentials, and, although it is obviously involved in the process of perception, it does not itself perceive; only the animal as a whole can do that.[36]

Another problem with the conventional view is the persistence of the idea that we see not the objects of the world before us, but only a picture in our brain or an image on our retina. Colin Blakemore again: "the subjects of seeing are not objects themselves, but the flat images of them which hide within the pupil of the eye."[37] As Peter Hacker suggests, "To argue that since we can see nothing without having a retinal image therefore what we see is the retinal image is like arguing that since we can buy nothing without money what we buy is money."[38] A combination of the above two misconceptions also explains why the upside-down retinal image is seen as a "problem": to worry about a flat, upside-down image is to assume that the brain can "see" this image, just as we can see the retinal image when we look into another person's eye, with the right kinds of instruments. But, of course, brains can't see anything, and "we" don't see our own retinal image either (because we could do so only by virtue of another little person in our head), so there is no reason to think that its being upside-down has any relevance at all (what does upside-down even mean, if there is nothing or no one looking at the image? Upside-down relative to what?).[39] The formation of a retinal image is simply incidental to what really matters as far as seeing is concerned—that an array of light has been reflected from objects in our visual field.[40]

There for the Taking

> [T]he senses can obtain information about objects without the intervention of an intellectual process.
> —*James Gibson*

Gibson's ecological theory stands in complete opposition to the conventional view of perception as an illusion. Gibson argued that perception starts not with the retinal image, but with the structure of light in the environment (the "ambient optic array": see below), which provides information to animals with a perceptual system capable of picking it up.[41] As animals are able to perceive this information directly—that is, without having to transform, enhance, or enrich it in some way—they act on the basis of what is in the environment, and not on the basis of a reconstruction inside their heads. In this way, Gibson rejected the conventional dualistic view of an "inner" mental world of perception by which we construct what the "outer" world looks like. For Gibson, there is just one world, in which animals can detect the information—the affordances—available and exploit them.[42] In Gibson's theory, then, the "problem" of the retinal image simply doesn't enter into things because this isn't the basis of perception. Instead, he identified another kind of problem that needs to be solved.

As Gibson pointed out, unless one is performing an experiment in a laboratory—where one can impose various kinds of stimulus energy on an animal's receptors in a controlled fashion—the intensity of light, sound, odor it encounters, and the things it can touch are highly variable from place to place and moment to moment as an animal moves about. The stimulation of its receptors, and the accompanying sensations, will similarly vary enormously. So, for Gibson, the big question of visual perception was: "How do humans and other animals obtain constant perceptions given that they are faced with such continual variability?"

Gibson's answer was to suggest that there are certain "higher-order" variables—"invariants"—present in the stimulus energy that do not change over time and place, despite the movements of the observing animal and the changes in the intensity of stimulation it receives. These invariants correspond to certain permanent properties of the environment (which is why they are invariant), and, as such, they constitute information about the environment that the organism can detect or "pick up."[43] For example, when you look at a rectangular table, you usually don't actually see it as a perfect rectangle because you can do that only if you look at it directly from above. Rather, you see a set of different and constantly varying trapezoid forms with different angles and proportions as projected to our moving point of observation. What doesn't change, however, is the relationship between the angles that sit diagonally across from each other (the cross-ratio), and these uniquely specify a rectangular surface (and also a rigid one).[44] Perception, then, is the activity by which animals and humans detect environmental invariants.

To get this point across, let's consider in more detail what Gibson refers to as the "optic array"[45] (most of Gibson's work was concerned with visual perception, but the same principles apply to other sensory modalities). Light rays travel through a transparent medium—air—and are reflected from the surfaces of objects. This light is available for a perceiver to use, providing its eyes are looking in the right place. The places at which light is available are termed points of observation (also called "station points"). At any of these points, light converges from all directions and forms three-dimensional angles (called "solid angles" to distinguish them from 2-D or plane angles) that are nested within each other at different scales (i.e., small solid angles are nested within ever larger ones), and these solid angles correspond to differences in the intensity of light. As the intensity and mixture of wavelengths of light from one angle is different from that coming from another, it forms a contrast. The arrangement of these contrasts is independent of the exact intensity or wavelength of the light that produces them; it is just the relative difference between them that matters. The structure or pattern made by these contrasts forms the optic array, and this is why light itself carries information: the structure of the optic array is determined by the kinds of surfaces, and their positioning in the environment, from which light has been reflected, and so the optic array is specific to a particular environment and (quite literally) reflects what it contains.

So, to put this in concrete terms, if you begin reading a newspaper on a sunny porch but move into a shady room as you get too hot what you find is that, as you move indoors, the paper doesn't suddenly change color to you, even though less light is now being reflected from the paper. The conventional view is that our nervous system corrects for the changing nature of the input, exchanging one illusory image of a newspaper for another, whereas the ecological approach suggests that we do not need to compensate internally in this way because the structure of the optic array itself remains unchanged despite the absolute change in light intensity (because it is the relative spatial patterning of the contrasts that matters, rather than their absolute values).

Action for Perception

The key to Gibson's theory is that animals must actively explore and attend to their environments to pick up the available information. This also means that, if environmental information should happen to become impoverished, so that perception suffers, an animal can take direct physical action to increase the quality of the information it gathers. Think of what you do when you want to see the label on a bottle that is turned away from you, or when you want to read a sign with very small writing: you turn the bottle toward you; you move closer to the sign. By moving around in the world, then, an organism transforms the optic array, and these transformations reveal the shapes, sizes, and locations of objects in the world. For Gibson, visual perception is not the reception of stimuli from the environment followed by the construction of internal representations, but an active sampling of the optic array that allows an animal to detect the information present in the world. This active sampling allows animals to perceive not only the "invariant structure" we described above, but also "perspective structure."[46]

When an animal moves and transforms the optical array, this provides information about its own locomotion—this is perspective structure. A flowing perspective structure indicates movement, whereas an arrested perspective structure indicates that the organism is at rest. In Gibson's theory, then, perception of the environment is always and simultaneously

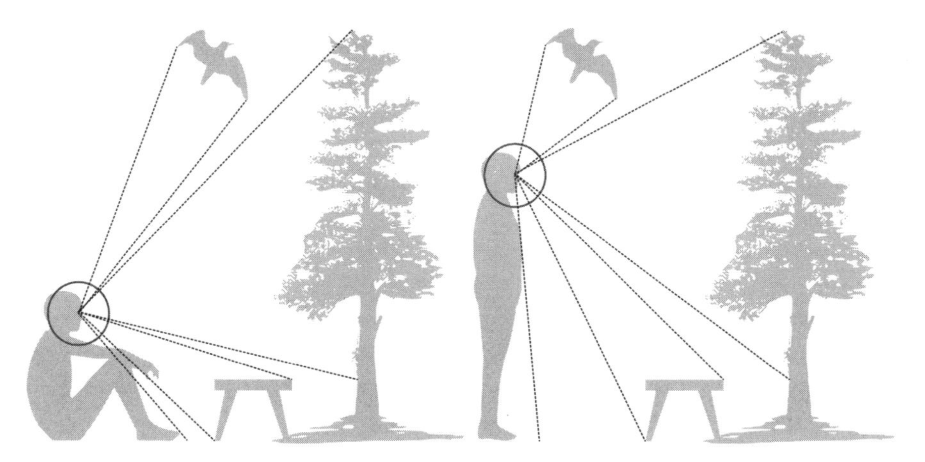

Figure 6.1. The optic array. Light is reflected from objects in the environment, and the "solid angles" so produced form contrasts that specify uniquely those objects.

a form of self-perception (nicely embedding the animal in its environment in a mutualistic way). Conversely, as we already noted, when an animal moves and transforms the optical array, there will be some aspects that do not change but remain constant over all transformations. This is the invariant structure that specifies the kinds of objects are present in the environment. This should now make clear where affordances come into play—they are the terms in which perceptual information is made available. That is, they are the invariants that are significant for a particular kind of organism—the ground's invariant of solidity affords walking for us, for example, while a wall's invariants of verticality and solidity afford leaning. We can bring in the umwelt here too, because, while the invariants are always present, they will matter more for some organisms than for others. A shoe affords protection of the foot to a human, but affords chewing to a dog: the invariants of shape are crucial for foot protection, but, obviously, they don't matter quite so much for chewing, whereas the invariants of resistance and texture may matter equally to both.

The most crucial point to take away from all this, however, is that the detection of invariant structure—and hence perception—is utterly dependent on the active manipulation of the optic array, so that the organism makes information available to itself. That is, perceptual information is available only to animals that are actively exploring in some or other way. We can link this back to our salticid spiders. If you recall, salticids can move their eye tubes, sweeping them over their cornea in complex patterns, and their bodily movements are guided by the detection of horizontal lines in their visual field. If we think of this in Gibsonian terms, the active sweeping of the eye tubes is the means by which the spiders detect the affordances—the horizontal invariants—of their environment and can then act on them accordingly.

Along similar lines, Alva Noë considers perception to be a form of "skilled access" to the world, in which animals are directly coupled to their environments.[47] Perception is not "in" us and it doesn't happen "to" us; it is something in which we actively participate. It is, to return to the beginning of the chapter, like a dance. Transforming the optical array so as to perceive invariants is also the strongest way to make the case that perception is a function of the mutual organism-environment relationship and can't be considered as something internal to the organism: whatever "cognition" is taking place, it is taking place not solely in the animal's head, but out in the world: action in the world can, justifiably, be considered to be just as "cognitive" as things that happen inside an animal's head.

Gibson and the "Denial of the Mental"

Arguments like this have led (quite frequently) to accusations that ecological psychology is "antirepresentationalist" or "antimentalist"[48] because Gibson argued strongly that there was always sufficient information in the optic array to specify the nature of the environment, relieving the organism of the need to internally process information.[49] On the one hand, this is a misrepresentation of Gibson's arguments. First, Gibson's theory was focused on perception and was never aimed at explaining other kinds of processes that (supposedly) involve representations. Gibson explicitly stated, in fact, that his theory "isn't to deny that reminiscence, expectation, imagination, fantasy and dreaming occur. It is only to deny that they have an essential role to play in perceiving."[50] Second, he said, again explicitly, that his theory "also admits the existence of internal loops more or less contained within the nervous system. . . . There is no doubt but what the brain alone can generate is experience of a sort."[51] What Gibson questioned was the usefulness of terms like "mental images" that people use to describe mental representations, mainly because it was unclear what such a term really meant: "we certainly do not conjure up pictures inside our head for they would have to be looked at by a little man in the head. . . . Moreover, the little man would have eyes in his head to see with, and then a still littler man and so ad infinitum."[52]

On the other hand, so what if Gibson's theory of perception is antimentalistic in the conventional "perception is an illusion" sense? Why should that automatically be regarded as a devastating criticism? After all, we have seen that there are conceptual problems with the conventional view, and there is a large body of empirical support for Gibson's theory.[53] We should also remember that mental representations are theoretical constructs that we use to try to better understand certain aspects of our own and other animals' lives. This doesn't necessarily make them "real," or even necessary to psychology (we discuss this more in chapter 10).

Finally, it simply doesn't follow that an argument in favor of direct, rather than indirect, perception is also an argument that everything happens outside the animal and nothing happens inside it (which is what these antimentalist arguments tend to imply). Gibson's argument was simply that the senses are not conduits by which "signals" or "messages" are sent to the brain, and that the brain is not a device that decodes and interprets these signals in order to construct static cognitive structures, or some kind of picture of the environment. In other words, there is nothing at all in Gibson's argument to suggest that we can or should exclude brains from the process of perception. Instead, it is an argument for changing the way we think about the brain. After all, if active exploration is the means by which perception is achieved, the brain simply must form an important part of that behavioral loop because—quite obviously—the brain is involved in controlling and orienting the perceptual organs in ways that permit information pickup.

In other words, it doesn't make sense to talk about "the brain" at all when discussing perception, in just the same way that it doesn't make sense to talk about "the eye" or "the ear."[54] Rather, we should think about the central nervous system as a fully integrated part of the means by which animals seek and extract information from the array of energy that surrounds them. In Gibson's view, the brain doesn't sit aloof from the senses, waiting to receive data on which it can put its inferential capacities to work. Instead, the central nervous system and perceptual systems "resonate" (metaphorically speaking) to the stimulus information in the environment. The analogy here is not one of constructing the world but tuning into it, as a radio receiver can be tuned to pick up certain frequencies (although, as Gibson notes, it must be a self-tuning receiver; otherwise we have the problem of the little man in the head, the homunculus).[55] The perceptual systems "hunt" for information until they achieve clarity, like picking up a radio station rather than noise, and this is self-reinforcing: the pickup of information reinforces exactly those exploratory actions of the perceptual organs that made the pickup possible, and the registering of information reinforces whatever neural activity in the central nervous system brings this about. This is very far from saying that "nothing" happens inside the organism in the ecological approach.

It is, therefore, much more accurate to understand Gibson's theory as an alternative model of cognition, broadly construed as how animals come to know their environments,[56] and not an anticognitive or noncognitive theory. Indeed (and somewhat ironically for the antimentalist critics), the lovely thing about Gibson's theory is that it is a theory of perception that is automatically a theory of cognition, with no false separation between the two.

Antimentalist criticisms of ecological psychology also tend to miss the important point that—regardless of where one stands on the issue of representation—the first step in any study of visual perception in any creature should be to determine how much information is present in the environment (or, if we are especially skeptical, whether there is any at all) before we begin to even consider formulating hypotheses about what's going on inside an organism's brain.[57] As Mark Rowlands—a philosopher and something of a champion of Gibson's views—points out, it is merely the sensible application of what he calls the "barking dog principle" (based on the old adage "Why buy a dog and then bark yourself?") as applied to evolved creatures. If there is information freely available in the environment, why would natural selection go to the trouble of building in internal mechanisms that do exactly the same job? A failure to consider the information available in the environment again runs the risk of assuming that the organism is performing a task in its head, when, in fact, the task is one that can be left completely to the world (more on which later). As should be clear, this isn't quite the same as saying that no internal activity takes place. Instead, it's an argument for giving the external environment as much attention as the inside of an animal's head when we are investigating their cognitive capacities.

The more precise and technical details of how animals directly perceive environmental structure, and how this contrasts with conventional psychological views, need not concern us here.[58] I have mentioned the difference simply because ecological psychology is often presented in pejorative terms (it certainly was when I went to university!) and described solely as a view that denies the need for any internal cognitive mechanisms at all. As should be clear, this isn't actually the case: ecological psychology is more of a reinterpretation of cognitive processes, acknowledging that these reflect the mutual interaction of an organism with its environment, so that things that happen in the world can be included as part and parcel of an investigation into cognitive mechanisms.[59] The more nuanced and accurate position is to say that representational systems or "ideas" are not mental phenomena alone, but are also ways of behaving and regulating our actions in the world: even we humans do not internalize "ideas" in our heads in a way that is completely divorced from reality (although formal schooling often makes it seem that way . . .); rather, we use these ideas to help regulate and control our encounters with our environment.[60] The only thing that Gibson actually denied was the entirely false separation of organism and environment, perception and action, that the conventional view entails, and it is this—and the concept of affordances—that we should keep uppermost in our minds as we continue.

CHAPTER 7

Metaphorical Mind Fields

Freud often compared the brain to hydraulic and electro-magnetic systems. Leibniz compared it to a mill, and I am told some of the ancient Greeks thought the brain functions like a catapult. At present, obviously, the metaphor is the digital computer.
—*John Searle*

The brain is said to use data, make hypotheses, make choices, and so on, as the mind was once said to have done. In a behavioristic account, it is a person who does these things.
—*B. F. Skinner*

So far, we've considered how our perceptual biases influence our tendency to anthropomorphize the world around us, and how, as big-brained mammals, we often fail to realize that much of the flexible ("intelligent") behavior that we see doesn't require very much in the way of a brain at all. We've also begun to explore some of the scientific biases that exist in psychology, and to see that alternative views are possible. In this chapter, we'll extend this argument and consider in more detail how one particular human bias, the one on which our scientific biases rest, may prevent us from appreciating what natural cognition is all about. Specifically, we'll consider the ways in which our scientific understanding of the world is structured by the use of metaphors, and how this has led to the dominant view of cognition as a brain-based process isolated from the world.[1]

What does it mean to say that we structure and understand our world through metaphor? In our everyday life, it refers to our tendency to understand certain abstract concepts in terms of other, more concrete, experiences. We understand the abstract notion of time, for example, by using spatial metaphors: we "look forward" to spring break, not least because we've "fallen behind" with our work and deadlines are looming, although we don't worry too much about this, as the "past is behind us." Similarly, we often think of our thoughts and ideas as food: we like something we can "get our teeth into," although we often find that we may have "bitten off more than we can chew."[2] As you may have noticed, we often use metaphors based on the bodily actions we can take in the world (moving through time; chewing on ideas), and, as we'll consider in more detail in the next chapter, this may well occur because most of our understanding of the world is grounded in—and built up from—our ability to act in it, so that even the most abstract of ideas (not excluding mathematical thought, according to some authors)[3] reflect what our bodies can physically achieve.

Using metaphors in this way doesn't mean that we literally believe that our thoughts are food, and that we will starve if we don't get any. Rather, we understand that a similar relationship exists between the equivalent elements in both the concrete and abstract domains and this is why we can draw the comparison: our thoughts can be "intellectually nourishing," if not literally so. In the same way, we draw analogies between items (i.e., interpret one thing in terms of another) by understanding the similarity of the relationship that exists between them: a bird's nest is like a human apartment (the relation of "home"), a dog wagging its tail is like a human's smile (the relation of "friendly behavior"). The ability to see beyond (another metaphor . . .) the juxtaposition of different elements to the relationship that exists between them (i.e., moving beyond observable features), so that it is possible to identify a similar relationship between an entirely different pair of items, is argued to be a key human trait—perhaps even unique[4]—and one that allows us to contemplate and understand the world in a more complex and abstract way than is available to many other creatures. Why, then, might this kind of reasoning lead us astray? Surely it is an extremely useful skill?

The answer is that it is, obviously a very useful ability, and, most pertinently, one that, as noted at the beginning of the chapter, plays a large role in scientific thinking. In science, we often have to deal with highly abstract concepts that would otherwise be very hard to grasp, precisely because they are so far outside our everyday experience. Metaphors are, therefore, an essential part of science.[5] One suggestion is that metaphors help us extend the boundaries of our knowledge (itself a metaphor . . .)

via a process of "catachresis"—which means to deliberately use a word or term to denote something for which, without the catachresis, there simply is no name. When we do this, we can bring into being entirely new ways of thinking, and pursue ideas that would never have occurred to us otherwise.[6]

So, this way, the structure of the atom was famously likened to the solar system, and DNA is often seen as a form of digital storage device. As we've already discussed in chapter 1, the action of natural selection is often compared to the intentions (the desires and beliefs) of humans. As the latter case also illustrated, however, this kind of reasoning can create problems when the metaphors employed are taken too literally. The same goes for the conventional view of perception discussed in the previous chapter: the suggestion that our brains "make inferences," "test hypotheses," and "present arguments" is, at base, metaphorical. As we noted, brains cannot literally do any of these things, but the misconstrual of this metaphor (or the simple failure to remain aware that a metaphor is being used) can lead one astray. No doubt all the neuroscientists referred to in the previous chapter would emphatically deny that they are arguing for a homunculus in the head, but by speaking of the brain as "inferring," "perceiving," and "asking questions," that is exactly what they are doing.

In what follows, we're going to consider a very powerful metaphor that helped shape the fields of psychology, cognitive science, and artificial intelligence for many years, and which may explain why we often get trapped into anthropocentric ways of thinking about the cognition of nonhuman animals. Specifically, we're going to consider the way in which many neuro-, cognitive, and comparative psychologists liken the brain to a computer (the "inferences" and "hypotheses" view of perception discussed in the previous chapter is obviously one aspect of this view). Indeed, some people go so far as to argue that the (human) brain is not just analogous to a computer in a strictly metaphorical sense, but that it actually is a computer that takes in input, processes it in various ways, and then produces a specific output.[7]

As with our anthropocentric tendencies, our use of the computer metaphor is so familiar and comfortable that we sometimes forget that we are dealing only with a metaphor, and that there may be other, equally interesting (and perhaps more appropriate) ways to think about brains and nervous systems and what they do. After all, given that our metaphors for the brain and mind have changed considerably over time, there's no reason to expect that, somehow, we've finally hit on the correct one, as opposed to the one that just reflects something about the times in which we live. Socrates considered the mind to be a wax tablet; John Locke, the seventeenth-century British philosopher, famously considered the mind to be "a blank slate," on which our "sense data" were written or painted; and, as the epigraph opening this chapter suggests, Freud compared the brain to a hydraulic system (with all its connotations of pressure build-ups and the need for "release"). The mind/brain has also been compared to an abbey, cathedral, aviary, theater, and warehouse, as well as a filing cabinet, clockwork mechanism, camera obscura, and phonograph, and also a railway network and telephone exchange. The use of a computer metaphor is simply the most recent in a long line of tropes that pick up on the most advanced and complex technology of the day.[8] This, in itself, should make us somewhat skeptical about claims for the computerlike nature of the brain; what should really make us wary, however, is how the computer metaphor took hold in the first place. To grasp this, we need to consider a little history.

Artificially Anthropocentric Intelligence

> Artificial intelligence is no match for natural stupidity.
> —*Anomymous*

> Chess is the *Drosophila* of artificial intelligence. However, computer chess developed much as genetics might have if the geneticists had concentrated their efforts starting in 1910 on breeding racing *Drosophila*. We would have some science, but mainly we would have very fast fruit flies.
> —*John McCarthy*

The computer metaphor first rose to prominence in the early 1950s. Prior to this, the telephone exchange served as our best metaphor for the brain. Brains were considered to be electronic switching devices that connected a stimulus to a response in the same way that a telephone operator connected one caller to another.[9] As the most prominent school of thought in psychology at the time was radical behaviorism, this analogy worked extremely well: for the most part (there were some exceptions),[10] behaviorists dealt not with internal, mental processes, but with brain-body behavior considered as a whole;[11] more specifically, their concerns were with behavior that could be controlled as a response to a stimulus, via learning. As it became clear that this stimulus-response account of the behaviorist approach couldn't provide an adequate account for all that an animal was (or wasn't) capable of learning, the idea began to gain ground that some internal processing had to mediate between a stimulus and a response.[12] At the same time that psychologists were rejecting and rethinking behaviorism, computer scientists were developing what came to be known as "artificial intelligence," and using computers to simulate cognitive processes. Psychologists began to cotton on to the idea that understanding brains and intelligence could be achieved not only via the analogy of the computer, but also by the actual use of computers to model and mimic the activities of the brain.

When Is a Turing Machine Not a Turing Machine?

The British mathematician Alan Turing, often regarded as the father of computer science, is widely credited with developing the "brain as computer" metaphor owing to his analyses of "Turing machines";[13] these were very basic devices—consisting of a read-write "head" (like that on a tape recorder) that could print, read, and erase symbols on an infinite tape of paper—that manipulated symbols in a very precise way. It is important to be aware that Turing machines do not actually exist; they are entirely abstract descriptions of a computing device that could be used to solve logical problems, via an "algorithm" (a set of rules followed in sequence).[14]

One can "build" an infinite variety of Turing machines, each of which is capable of computing a single specific sequence of numbers depending on how its read-write head interacts with the symbols on the tape (i.e., based on its specific algorithm). Building on this idea, Turing proposed that it was possible to develop a "universal" Turing machine—one that would be able to simulate the operation of any other possible Turing machine—so that, instead of being able to compute only a single

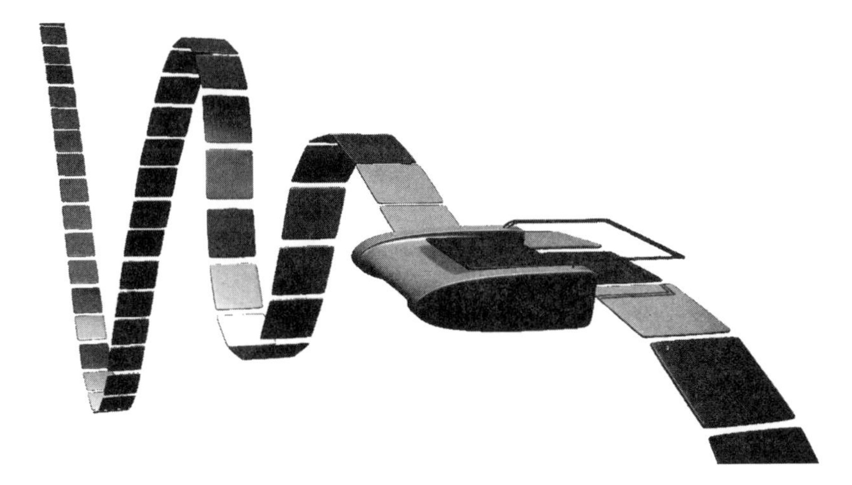

Figure 7.1. A Turing machine is an abstract device thought up by the British mathematician Alan Turing to demonstrate how numbers can be computed through the use of an algorithm (a set of rules followed in sequence).

sequence of numbers, a universal machine would be able to calculate any possible sequence of numbers, provided there was a specific Turing machine whose operations it could reproduce. Strange as it may seem, it was proved that this purely mechanical procedure—the "algorithm" used by Turing machines—could be employed to calculate the answer to any question that any other kind of computer could calculate (that is, not just mathematical questions, but any kind of question at all, provided it could be encoded by the symbols used by the Turing machine). This led to much excitement and speculation that perhaps human thought was a similar kind of algorithmic, symbol-manipulating process, and, even more excitingly, perhaps the brain was a real-world universal Turing machine.[15] This opened up the possibility for modeling human thought, language, perception, categorization—whatever process one liked—using a digital computer. Why? Because, like a universal Turing machine, and supposedly like the human brain, computers use algorithms to perform calculations (or computations).

As a result—and as computers became a reality, and not just theoretical proposals—it seemed possible that humans would be able to create a brain capable of humanlike thought using human-made silicon chips, instead of biologically evolved neurons.[16] This follows logically because the computational processes of a Turing machine do not depend on the actual materials that are used to make it: a Turing machine can be made out of anything you like, not just paper tape and magnetic heads, but anything at all, from "two kinds of pebbles and a roll of toilet paper," as Jerry Fodor once put it,[17] to "cats, mice and cheese," as the philosopher Ned Block once suggested.[18] Moreover, because the brain alone was seen as the key to understanding cognition—based on the idea discussed in the previous chapter that, in order for us to perceive and think about the world, representations of that world must be constructed to compensate for the poor quality of the information received by the sense receptors—it meant that bodies and the environment became completely irrelevant to the study of cognition. This further reinforced the idea that it was possible to create humanlike intelligence in a computer; a computer can be considered equivalent to a brain, but not to an active, moving body.

From this point on, psychological processes—in both human and nonhuman animals—became closely identified with various kinds of "information processing." The idea was that sensory input came into the cognitive system; the cognitive system algorithmically manipulated symbols,[19] as would a Turing machine/digital computer, and then produced an output that manipulated the body. It is at this point that the clear separation of perception, cognition, and action, which we have noted in earlier chapters, began to be made, and efforts to understand the workings of the mind (and "thought" and "intelligence") came to mean efforts to identify and understand the "information processing" that occurred between sensory input and motor output. With the computational metaphor in place, it became almost inevitable that the brain would be seen as the equivalent of computer hardware, with cognitive processes operating like the brain's software: an idea that has permeated modern Western culture at all levels. In the film *The Matrix*, for example, it was possible to download computer programs directly into people's brains via a portal at the back of the head, obviating a long-drawn-out learning process and providing the recipient with expert abilities in, among other things, kung fu (again emphasizing that the body is largely irrelevant to the development of even such highly physical skills; a dangerous assumption, as we shall see in chapters 9 and 10).

Aside from the problems of neglecting bodies and environment, another problem of the development and use of the "brain as computer" metaphor is—as Andrew Wells points out in his marvelous book on the subject—that it completely misrepresents a Turing machine and also Turing's aim in developing them.[20] To get the full story, and so understand what Turing was attempting, you should really stop here and read Wells's book for yourself, but assuming you don't do that (even though you should), a brief summary will serve our purposes.

When Turing's paper was published, way back in 1936, a "computer" was not a machine, but a person. A person who computed sums. Turing's aim was to try to find a way of mechanizing this process, thereby producing a labor-saving device that could do the work of human computers. As we noted above, Turing conceived of his machine as an infinite paper tape, divided into squares on which symbols could be read and printed. This tape passed through a head that could move either to left or to right, one square at a time, and this head could both read what was written on the tape and print on it. In most books and articles in which a Turing machine is discussed, this whole kit and kaboodle is used as an analogy for the mind or for cognitive processes: inside our heads, it is argued, we have a Turing machine that receives input in symbolic form, manipulates it, and then provides an output. The tape of a Turing machine is, in essence, a model for human memory.[21]

Now, the truth of the matter is that this couldn't be further from what Turing was actually attempting to model. Remember that he was trying to conceive of a machine that could calculate sums in the same way that a human computer calculated them. How do we calculate sums? If they're long and complicated, most of us do it on a piece of paper—maybe even graph paper—using a pen or pencil. In this light, let's consider the abstract Turing machine again. The paper tape that is usually seen as internal memory was, for Turing, part of the environment. Specifically, it represented the paper on which a human computer could work out his or her sum. So the paper tape is not a model of memory in the head but a model of graph paper in the environment.[22] Equally, the "machine-head" that reads and writes the symbols does not represent the cognitive processes taking place inside a person's brain, but instead represents the person as a whole, using pen and paper to calculate sums. Wells refers to this setup as a "mini-mind" to get across the idea that these can be either complete descriptions of very simple minds or, alternatively, partial descriptions of more complicated minds (after all, there is more to a person than simply computing sums). So a Turing machine actually consists of a mini-mind that has a finite number of states (because human memory has finite limits), and an infinite tape divided into squares (because, when real people do real calculations, their ability to do so is not usually limited by their having access to only a fixed amount of paper). The combination of the state of the mini-mind and the contents of the tape is called a "configuration." The current configuration determines the moves the machine makes, what it prints, and what the succeeding configuration will be.

It couldn't be clearer from this description, then, that the symbols a Turing machine manipulates are outside the mind, and not part of it.[23] A Turing machine is, as a result, a very ecological contraption, in Gibson's sense of the word. Computing is about the relationship between a human computer and his or her environment (which consists of the paper and pencil used to do the sums). One cannot understand the behavior of a Turing machine simply by looking at the state of the mini-mind (the person, if you like), nor can one understand what the computer will do just by looking at the tape (the environment). To understand a Turing machine's behavior, one has to look at the relation between the agent and the environment. Wells uses this insight to argue that a mind is both formed and maintained by the continuous interactions between an agent and the environment.[24] Turing modeled exactly these kinds of interactions, but only in a very specific context. It was never his intention to provide a general analysis of human behavior, nor to suggest that all human cognition conformed to this specific kind of computational process. Indeed, Turing's concerns were clearly mathematical, rather than psychological. He was simply interested in what numbers it was possible to compute, as a human did, using a pencil and paper.[25]

So, if Turing's machines were never intended to be a model of the mind or of mental processes, where did our current idea of the brain as a computer come from? For the answer, we have to cross the Atlantic. The first real-world version of a Turing machine was constructed for the United States army and known as the ENIAC (Electronic Numerical Integrator and Computer). Owing to the way it was built, and the fact that it was a special-purpose Turing machine (rather than a "universal" one), the computer's entire physical hardware had to be changed and rewired every time a new kind of calculation was needed. In the late 1940s, John von Neumann was one of several people charged with the task of making the ENIAC more convenient and useful, and it was he who designed the architecture used by all modern computers today: a central processing unit, a main memory, a set of peripherals (like keyboard and monitor), and a second memory that could be used to store information externally, like hard drives, CDs, and memory sticks. It is, therefore, to von Neumann that we owe the "brain as computer" metaphor, as it was he who helped create self-contained digital computers. In addition, it was he who specifically compared his computer architecture to that of the brain, suggesting that the central control (CPU) of his computer corresponded to the "associative" neurons of the human nervous system, and that the input and output devices were the equivalents of sensory and motor neurons, respectively.[26] This "von Neumann architecture" is one that has been used in many different kinds of artificial intelligence projects and programs, and it is this, rather than a universal Turing machine, on which our metaphors of mind are based. Our notion that Turing machines represent the basis for our current view of cognition is completely off-track.

I mention all this here, of course, to highlight the possibility that, had people recognized the true psychological implications of Turing machines (that they reflect the ongoing mutual relationship between a "computer" and her environment, and are not a model of the mind divorced from the environment), we might have had a very different view of cognition and the brain, and a different kind of psychology might have been the result. Indeed, this is Wells's point: he explicitly shows how one can marry Gibson's ecological theory with Turing's theory of computation to provide a formal model of affordances (one that works better than the available alternatives),[27] and one that can serve as an alternative model of cognition. Space doesn't permit a detailed examination of Wells's argument here, but, in essence, affordances can be characterized and studied as the "configurations" of a Turing machine (the state of the "mini-mind" and the contents of the tape), an idea that captures the complementarity between animal and environment that is essential to Gibson's theory.[28] Just as affordances "point" both ways—toward the animal and the environment—so do the configurations of a Turing machine. The Turing machine model also gets at the issue of internal structure in the organism versus external structure in the environment, which has been the source of much of the criticism of Gibson's model. As Wells notes, there is a theoretical trade-off between internal and external structure in a Turing machine: a machine with only two possible internal states can compute numbers provided it has access to an external alphabet that is large enough. Equally, a Turing machine that has access to only a two-symbol alphabet can compute numbers provided it has a large number of possible internal states. The Turing machine model suggests, therefore, that structure in the animal will complement structure in the environment.[29] This means that one cannot simply assume a particular behavior results only from structure in the animal versus that of the environment or vice versa: it should reflect a trade-off between these two, and to discover what that is, you have to go and find out (the point we made in the previous chapter).

Finally, the idea of a universal Turing machine, once properly understood, also tends to support a Gibsonian view of the world. Unlike those Turing machines that compute only a specific sequence of numbers, and which always start on a blank tape, the universal machine works by beginning on a tape that already contains a string of symbols, which allows it to produce the output of the machine it is simulating. As the tape is actually part of the environment, a universal machine supports the notion that the information available for perception is found mainly in the environment, and not in the head.[30] Wells's combining of Turing's theory with Gibson's theory is, in a way, wonderfully subversive, because it brings together the most cognitive of all models in psychology—the Turing machine as isolated brain—and marries it to a theory that requires complete complementarity between organism and environment. Of course, in another way, it is not subversive at all, because it is merely correcting the misconception that the notion of a Turing machine supports the "brain as computer" metaphor that currently holds such sway.

Alternative Metaphors for the Brain?

Looking at Turing machines from an ecological perspective, and highlighting the differences between Turing machines and von Neuman architecture is a point well worth making because, although the computer analogy built on von Neumann architecture has been useful in a number of ways, and there is also no doubt that work in classic artificial intelligence (or, as it's often known, Good Old Fashioned AI: GOFAI)[31] has had its successes, these have been somewhat limited, at least from our perspective here as students of cognitive evolution.

As a number of cognitive scientists and roboticists have pointed out over the years,[32] the classical AI perspective, with its emphasis on the algorithmic manipulation of symbols using a von Neumann architecture, naturally gravitated toward those aspects of cognition, like natural language, formal reasoning, planning, mathematics, and playing chess, in which the processing of abstract symbols in a logical fashion was most apparent. As a result, classic artificial intelligence also placed humans front and center, with the focus of research resting squarely on understanding some peculiarly human aspects of intelligence: none of them are very athletic—they don't require an active organism in the Gibsonian-sensorimotor sense—and none of them require any specific interaction with the environment, as opposed to seeing the environment simply as the arena in which the products of these computations are played out. Unfortunately, this rather arbitrary initial emphasis on these particular (and specialized) kinds of logical, algorithmically based tasks gained such momentum that researchers came to the conclusion that everything brains did (human and nonhuman alike) was simply a form of logical reasoning, and that they employed an algorithmic process to achieve this. I say "unfortunately" because, while this view (eventually) managed to generate a computer that could beat the world chess champion, it has, so far, failed to give us any real insight into the mechanisms that underlie the more natural forms of intelligence we've been discussing up to now: how adaptive behavior is produced in a changeable environment. In human terms, this would include things like how we recognize a face in a crowd, how we coordinate our movements and manipulate all the objects necessary to make cup of tea, or even something as apparently simple as how we manage to walk, run, and even hop over uneven ground without falling flat on our faces.

You should now begin to see the problem. Our metaphor of the brain—and hence of cognitive processes—is one that was originally derived from a heavily anthropocentric focus on a few peculiar human cognitive achievements, all of which involved abstract symbol manipulation. As we've now seen, this in itself was derived from a misreading of Turing's work on computable numbers: work that made no claims of generality as far as psychology and cognition were concerned, but dealt only with a very specific human activity (so one cannot accuse Turing of misplaced anthropocentrism; he was quite clear about the aim of his work). As Wells shows, when properly understood, the Turing machine model can be seen as a critique of current cognitive approaches, one that supports the underlying philosophy of ecological psychology[33] (and, as we shall see later, the ideas of "embodied" and "distributed" cognition). Although it is true that certain aspects of our cognition can be understood and analyzed as computational processes involving the manipulation of symbolic representations—or, as some would suggest, are best understood via this kind of analogy—you should now appreciate that this isn't quite the whole story: not for us, and certainly not for other, nonlinguistic species.

As we've noted in our consideration of both Gibson's and Turing's work, the missing ingredient in all this is the recognition that the body and the environment really do matter as far as cognition goes. After all, when brains evolved initially, they did so in animals that already possessed bodies, and long before they possessed anything that we would recognize as a brain.[34] By failing to account for this (and indeed by completely misinterpreting the nature of a Turing machine itself) the computer metaphor has generated a view of cognition as a process that has no real link to the body or the outside world, taking place purely in the brain alone.

What is worse is that we have taken this strange view of cognition—that it takes place inside the "Turing machine" of the brain and involves the disembodied, logical manipulation of internal representations—and applied it directly to other animals. The metaphor of the computer and the idea of a computational, representational mind is one that pervades studies of comparative cognition[35] (even those articles that are critical of the more anthropocentric/anthropomorphic interpretations of such studies nevertheless take the existence of von Neumann–like computational, representational processes to be axiomatic, rather than an assumption to be tested).[36] What is also interesting is that we have applied the computational model to other animals even though it doesn't adequately explain most facets of our own natural cognition (leaving aside the fact that this kind of psychological generality was never the intention of the earliest proponents of this model in the first place).[37]

By presenting an essentially disembodied view of cognition, the computer metaphor, with its input-output (stimulus-response) structure, also suggests a very static picture of animal life. Like the computer on your desk, an animal just has to sit there until a stimulus (an input) comes along and causes it to act. As we have seen, the vast majority of animals are not passive in this sense: they seek out relevant and significant resources in the world, in an active, animate fashion. This again is a consequence of their possessing a body and not only a brain (and of possessing bodies before brains). To ignore the body and the environment when considering how animals behave in an "intelligent" fashion is, at the very least, to miss out on half the story. Indeed, we've already seen how the ears of the cricket and the eyes of the *Portia* spiders are highly relevant factors to consider when we are trying to understand the behavior of these animals in their natural environments. As we'll discover more fully in the following chapters, we are likely to gain a better understanding of the natural kinds of intelligence that we see every day (and engage in ourselves) only when we take the body as seriously as we do the brain.

Returning to the Age of Steam

> Steam is no stronger now than it was a hundred years ago,
> but it is put to better use.
> —*Ralph Waldo Emerson*

Given these problems with the "brain as computer analogy," how, then, should we think about cognitive processes? One solution, as we've seen, is the "ecological" computational approach suggested by Andrew Wells. We can also, however, consider other kinds of models besides these computer-based metaphors. Or, as Tim van Gelder, a philosopher at Melbourne University in Australia puts it, "What might cognition be, if not computation?"[38] His suggestion follows on from that of the ecological psychologists, to some degree, by recognizing the dynamic way an animal's sensory systems interact with its motor systems and how these interact with the world. As the nervous system, body, and environment are simultaneously changing and influencing each other in a continual cycle of adjustment (they are "dynamically coupled"), we should properly consider a "cognitive system" to be a single, unified system that encompasses all three elements and doesn't privilege the brain alone (especially a disembodied and autonomous one, translating abstract input relations into similarly abstract output relations).[39] Interestingly, this notion of dynamic coupling, where each change in one element of a system continually influences every other element's direction of change, can be captured in another machine-based analogy. As van Gelder suggests, a better model for how cognition works may be not a modern digital computer, but something like a Watt governor.

A Watt governor, also known as a flyball or centrifugal governor, is a device used to regulate the speed of a steam engine, regardless of changes in the workload of the engine or the fuel supply. It was named after James Watt, who designed some for use on the first steam engines (although it should be noted that Watt himself didn't invent the governor: governors of a similar design had already been in use in windmills for many years). The governor consists of two flyballs (hence the name) connected to a spindle by two flyball arms. The spindle is attached directly to the shaft of the steam engine. If the speed of the spindle increases, the flyball arms move upward owing to centrifugal force. The clever bit comes in here: the flyball arms are connected to a throttle valve that regulates the amount of steam that enters the engine. When the engine speeds up, the upward motion of the flyball arms closes the throttle valve, thereby reducing the steam input to the engine and slowing it down. Of course, as the speed of the engine falls, so, too, does the spindle, which means that the flyball arms drop. This has the effect of opening up the throttle valve, which allows more steam into the engine, which then speeds up.

As a consequence of this constant adjustment of the spindle, flyball arms, and throttle valve, the engine maintains a constant speed through smooth and swift adjustment, despite fluctuations in the steam pressure and workload. It should be apparent that, despite the way I've described it above, it is very difficult to identify a discrete sequence of events in a flyball governor because everything is happening continuously and smoothly, all at the same time: the angle of the flyball arm determines the speed of the engine, but, of course, it is the speed of the engine that determines the angle of the flyball arms: the angle of the flyball arms and engine speed are both determining, and determined by, each other. The Watt governor therefore solves the problem of constant engine speed in an entirely noncomputational, nonrepresentational way.

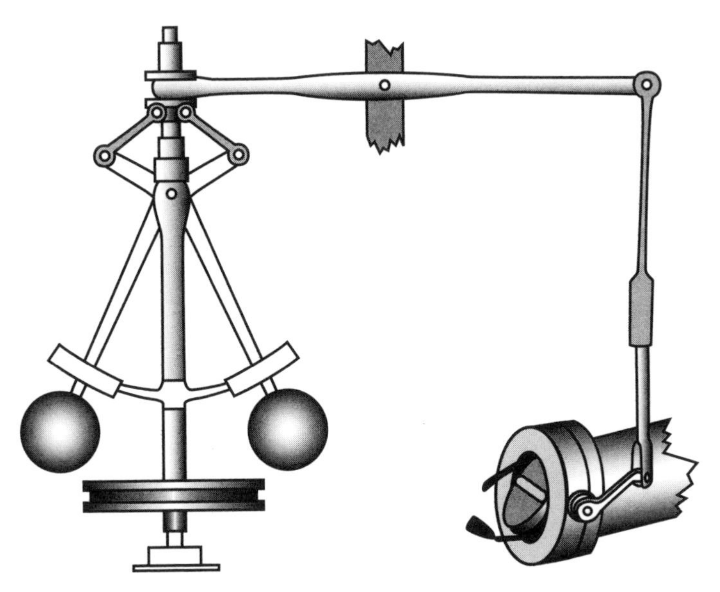

Figure 7.2. A Watt governor for a steam engine. The flyball arms are connected to a throttle valve that controls the amount of steam entering the engine and a spindle that is connected to the drive shaft of the engine.

Of course, as van Gelder (1995) argues, one could, in principle, come up with a computational, representational governor that would do much the same job. Van Gelder himself produced an example of just such a computational algorithm:

1. Measure the speed of the engine.
2. Compare this speed against the desired speed.
3. If no discrepancy was detected, then return to step 1. Otherwise:
   a. Measure current steam pressure.
   b. Calculate desired alteration in steam pressure.
   c. Calculate necessary throttle valve adjustment.
4. Make throttle valve adjustment.
5. Return to step 1.

What we need to remember here, however, is that this computational solution, while fine in principle, isn't the one that actually solved the problem of controlling engine speed, at least partly because the technology needed to implement such a computational approach wasn't available at the time. However, it is also noteworthy that the governor that was invented did its job superbly well. It wasn't just an inferior and primitive fix, the best that could be done in the absence of computer technology. It is also true that the computational solution is a lot more complicated in terms of the parts needed and operations performed. There have to be devices that can measure the relevant parameters, as well as devices to implement the response needed. A computational governor is therefore likely to be more expensive to build and run than the Watt governor, and there are more parts that can go wrong. This is relevant from our evolutionary perspective, since evolution is a thrifty process and tends toward the cheapest possible route to solve a problem effectively. A computational solution is not the only possible way to solve the problem of variable engine speed, as the Watt governor demonstrates, and we should take this lesson to heart: just because one can very easily come up with computational solutions to problems, including those of animal cognition, we should not be misled into thinking these are the only solutions possible; other, potentially cheaper, equally effective, solutions may be there for the asking. The noncomputational solution to the problem of engine speed is in no way inferior to the computational one; it is merely different, but it does the job as well, if not better, and at a lower cost.

Still, even if we agree that the Watt governor is just fine, and there is no necessity to replace it with a fully computational algorithmic device, one could still make a case that the Watt governor itself is, in fact, using representations and is, therefore, a computational device. One could argue that the angle of the flyball arms does, in fact, "represent" the speed of the engine because the angle of the flyball arms is correlated to engine speed. One could, in principle, use this angle to stand in for how fast the engine is running.[40] This does, however, miss a very important point about how the governor does its job: although there is indeed a correlation between angle arm and engine speed, the angle of the arms is at all times determining the amount of steam that can enter the engine, and hence at all times the speed of the engine depends on the arm angle, just as much as arm angle depends on engine speed. To argue that one "represents" the other is massively oversimplistic, and it also fails to capture the fact that the system is dynamic and in a constant state of flux. Consequently, if the governor is not strictly representational, then it can't be computational (if we stick with the definition that computation involves the rule-determined manipulation of representations). Without representations, and because of the mutually determining nature of each element in the governor, one cannot identify any discrete algorithmic steps in the operation of the governor, and there is a sense in which the system simply cannot be considered computational (but we will revisit this below in slightly different terms). The cut-out-and-keep message here, then, is that cognition need not be—either by definition or by logical inference—a purely computational process. It also suggests strongly that flexible, intelligent systems need not be separated into "hardware" and "software" components (the sticky "wetware" of the brain and its cognitive processes)—they are one and the same. In other words, although the computer metaphor has been, and perhaps still is, useful in helping to predict and explain certain aspects of (human) psychology, we shouldn't make the mistake of thinking that this means that natural cognition really is computational and therefore that the brain really is some kind of biological computer.

Timing Is (Almost) Everything

> Observe due measure, for right timing is in all things the
> most important factor.
> —*Hesiod*

Of course, cognitive systems are no more like Watt governors than a brain is like a computer—not literally. As we've noted, both are merely metaphors. The computer metaphor has, however, been taken both very literally and very seriously, and has promoted a very particular view of cognition that has been widely and wholeheartedly adopted by many researchers.[41] Looking to the Watt governor as an alternative metaphor is useful not because there is any suggestion that cognitive systems actually work in this way, but because cognitive systems may be better understood as "dynamical systems" where inputs, internal processes, and outputs—or, to put it more concrete terms, the environment, the brain, and the body's actions—are coupled like the spindle, angle arm, and throttle valve of the governor. Dynamical systems present us with more useful means for understanding and thinking about physically embodied, environmentally embedded organisms than do standard computational models.

I want to expand on this point a little further because, as our discussion of Turing machines above makes clear, if we define a dynamical system as one that shows state-dependent change (i.e., the future state of the system depends causally on the current state of the system), then computational systems are, by definition, dynamical systems.[42] In a Turing machine, the future state of the tape depends causally on the current state of the head and what is currently written on the tape, and this represents the coupling of the mini-mind with the tape environment. Looked at in this way, computational systems can be seen as a specific subset of the kind of dynamical systems that includes the Watt governor.[43] This inclusive definition means we can account for all cognitive processes using a dynamical systems approach (potentially anyway—we are actually nowhere near doing so), without being forced into a situation where we're trying to explain how two very different sorts of processes—computational and dynamical—came into being, and how they fit together.

So far, so good. But if computational systems are dynamical systems, what, then, is the real difference between these kinds of systems and a Watt governor? Michael Wheeler identifies at least two factors that seem to be key in differentiating between them.[44] First, computational systems, by definition, involve the use of representations: to do their job they must access, manipulate, and transform symbols. As we noted above, if one felt really strongly about it, one could make a case for a representational version of a Watt governor, but we also showed that representations were not essential to get the job done (that was the whole point of the example). So that's the first difference: a computational dynamical system absolutely requires representations, whereas a noncomputational dynamical system does not.

The second difference is more important, and, much like good comedy, it's about timing. In a computational system, time is reduced to the mere sequencing of events; in a Turing machine, things have to happen in the right order, but the time it takes for transitions between states to occur is not dealt with at all, and there is no specific theoretical reason why things should happen in a specified amount of time. Similarly, the amount of time the machine should remain in a given state is not considered, because, in a Turing machine, this would serve no specific function. Time simply doesn't matter. Of course, in the real world, one could argue that computational events have to occur swiftly enough to enable the problem to be solved in good time, but outside of that, time has no role to play.

As we saw in the Watt governor example, this isn't true of noncomputational dynamical systems. Instead, they exhibit "richly temporal phenomena."[45] This means simply that the actual rates and rhythms that characterize a particular process play an important and central role in getting the job done. This could be the way that the underlying physical processes of the brain work (how long it takes for a neurotransmitter, like nitric oxide or glutamate, to diffuse through the brain, for example, or how long it takes for such neurotransmitters to modulate neuronal activity), which in turn could affect the specific durations or rates of change in other physiological processes. Similar intrinsic rhythms in the body may also be important, as will other aspects of bodily dynamics that relate to, for example, the mechanical properties of muscle, which dictate where and how fast an animal can move. These bodily processes may, in turn, need to be synchronized precisely with temporal processes occurring outside of the animal in the environment.

This issue of timing is very clear in our Watt governor example, where the coupling of the different parts, and the specific rhythm and timing they displayed, were crucial to its success in controlling engine speed. Interestingly, when more sophisticated governors were developed, they showed behavior that was much less effective than that of the earlier models (which is rather counterintuitive): the new "improved" models "hunted" for a steady speed, continually speeding up and slowing down, rather than smoothly maintaining a steady state. This was because superior manufacture of the component parts meant that they generated less friction, and this, in turn, meant that speed adjustments were effected much more quickly. Greater friction in the older models meant that any changes in engine speed took longer to feed though the system, and this intrinsic quality helped the governor to perform the job at hand more effectively. Of course, friction and heat are features of computational systems as well (this is why your computer has a fan built into it), but the point is that, in a computational system, these are merely problems to be overcome by engineers, not an integral part of the computational process.

Thinking in terms of dynamical systems with a "rich temporality" also provides us with a new way of viewing the "failure" of evolved knowledge in the face of environmental change. Returning to our digger wasps, we can see their routine—preparing a chamber, inserting a bee into it, and laying their eggs—as a dynamical interaction of the wasp's internal state (of readiness for egg laying), its actions in the world (hunting and chamber preparation), and its environment (the presence of the chamber, the proximity of the bee to the entrance). The "failure" of the wasp to begin its "routine" in the middle is a failure only if we assume there is an underlying algorithm being followed. If, instead, we consider that the wasp's brain and body are making continual adjustments to an environment that is continually being changed by the presence of the wasp (and so changing the wasp's state at the same time as the wasp changes the state of the environment), we're less inclined to see a failure and more aware of the fact that we're watching a dynamically coupled system in action.

One must be somewhat cautious in adopting a more dynamical approach, however. In particular, the philosopher of cognitive science Andy Clark has noted that, because dynamical systems approaches are concerned with the state of a system as a whole—so-called total state explanations—we can potentially lose as much as we gain from adopting this approach over the computational approach. Clark's argument is that a dynamical approach obscures the "intelligence-based" route to evolutionary success that characterizes living cognitive systems, as compared to the other kinds of physical dynamical systems that exist in the world, such as river flow systems.[46]

As we noted in chapter 5, brains evolved in order to allow animals greater control over their environments and their destinies. Although we have spent a lot of time in this chapter putting the brain in its place, it would be foolish to suggest that brains don't matter. Brains are crucial as a location of behaviorally relevant activity, and this, as Clark notes, must mean that brain-involving dynamical systems are very different from other kinds of dynamic physical systems.[47] Brain-based systems achieve the kinds of behavioral flexibility that we're interested in precisely because the brain is able to alter the "information flow" through the system cheaply and in a wide variety of ways. If we deal only with the overall state of the cognitive system, then those aspects of how information flow is specifically channeled and directed by the brain get lost. If, however, we are mindful of this possibility, and we consider the inner flow of information within the brain as seriously as we do the overall state of the system, we can generate what Clark calls a "powerful and interesting hybrid: a kind of dynamical computationalism." By this, he means we could combine the "standard" computational and information-processing concepts with the coupling and richly temporal phenomena of truly dynamical systems.

His suggestion is that, rather than treating computational systems as fundamentally different from noncomputational ones as described above, we should attempt to combine the two so that the conventional computational approach is given a new dynamical dimension. His argument, then, is to take Wheeler's idea of computational systems as a specific subset of dynamical systems but to try to erode the distinction between them by allowing richly temporal phenomena to transform the standard computational approach. This may well be a productive way forward: as the complexity of sensory, motor, and physiological systems increases, and more complex behavior is possible, then, as we mentioned earlier, one would predict that the brain would have to be more strongly involved in altering information flow through the brain-body system in order to provide the kinds of temporal coordination needed to permit temporally rich adaptive behavior to emerge.

With this caveat in place, a dynamical systems approach, with its emphasis on rates, rhythms, and synchrony, is preferable because it is one that, by definition, naturally gives body and world their due when it comes to cognitive processes because, as Wheeler makes clear, these nonneural components will also act as pacemakers and rhythm-setters in causally important ways, in conjunction with those taking place in the brain.[48] Even better, perhaps, a dynamical systems approach treats the brain as an integral part of the body, and not as the all-powerful highly privileged computer that "tells" the body what to do. Like nonneural bodily processes, the neural activity of the brain has its own intrinsic rhythms and undergoes change at different rates. These, in turn, must be synchronized with the events that are happening in the body and the environment to produce effective behavior. The standard computational model, which keeps perception, action, and cognition as separate, independent processes, and (implicitly) assumes these need occur only in sequence, but not in real time, is both fundamentally "disembodied" (because cognition does not depend on any aspect of an animal's intrinsic physicality) and "disembedded" (because the environment plays no intrinsic role in helping to regulate the cognitive system, but is merely the "stage" on which the products of a disembodied cognitive process are played out). We want a richly rhythmic time-dependent view that accords with the lives of real, richly rhythmic time-dependent animals, and so we shall continue to pursue a dynamical approach in the next chapter.

Chapter 8

There Is No Such Thing as a Naked Brain

You've got the brain of a four-year-old boy, and I bet he was glad to be rid of it.
—*Groucho Marx*

We can discover more about the dynamical approach to animal cognition and behavior by moving away from the more abstract systems of the last chapter, and taking a look at real brains, and the ways in which they are coupled to the environment. Walter Freeman, a neurophysiologist at Berkeley, has spent the last thirty or so years performing intricate and meticulous experiments on smell, vision, touch, and hearing in rabbits (mainly) and has worked out a model of learning based on the kind of dynamic coupling between brain and environment suggested by the dynamic systems approach.[1] Before we can go into detail about Freeman's work, however, we first need to cover a little more ground on the theory behind dynamical systems, so that we can more fully appreciate Freeman's views on how brains, bodies, and the environment fit together.

Mathematically speaking, a dynamical system consists of a number of "state variables" (e.g., the engine speed and flyball arm angle in the Watt governor) that specify the state of the system at a given time, along with a set of equations that describe how those variables change over time. There can also be certain values that specify quantities that can change the state of the system, but aren't themselves changed as a result: these are called the parameters of the system. Putting everything in these terms allows us to think of a dynamical system as a form of graph—a multidimensional "phase space"—where the number of dimensions is set by the number of state variables of the system. In such a phase space, each possible state of the system (all the possible combinations of all the state variables) can be represented by a single point. Changes in the state of the system over time can then be plotted as a curve in the phase space, called the "trajectory" of the system. As in all things, a concrete example of a dynamical system will make this notion of phase spaces and trajectories easier to grasp, but instead of returning to the flyball governor once again, we'll use Rolf Pfeifer and Josh Bongard's example of "Puppy," an ingenious robotic dog developed by Japanese roboticist Fumiya Iida, to do the job.[2]

Puppy is a four-legged running robot, with a total of twelve joints (one at each hip and shoulder, one at each knee, and one at each ankle) with springs that connect the lower and upper parts of each leg. There are also pressure sensors on its feet that indicate when its foot is in contact with the ground. The control system of the robot is extremely simple: there are motors that move the shoulders and hips backward and forward in a rhythmic fashion. That's it. If you place Puppy on the ground, it will scrabble around for a bit, as it gains purchase on the surface, and then settle into a running gait. This is due to the tightly coupled interaction of the control movements of its hip and shoulder joints, other aspects of its anatomy (its overall shape and how the springs are attached), and the environment (the friction on its feet produced by the ground surface and, of course, the force of gravity). With this information to hand, we can now look at Puppy's behavior in terms of the general features of dynamical systems that we identified above. So, to follow Pfeifer and Bongard's example, we can take the joint angles of the legs as our state variable, and we can then characterize and so capture Puppy's movements by looking at how these change over time (their trajectory through phase space). As there are two joints per leg (knee and ankle), the phase space will have eight dimensions, where each point in that space represents a set of values for all eight joints. Points that are close to each other in space represent similar values of the joint angles (and so similar ways of moving), and those far away from each other characterize very different values of the joint angles (and therefore very different ways of moving, e.g., walking versus running). When Puppy moves, the joint angles change continuously, and the point that represents the value of the joint angles at a particular time moves across phase space accordingly, plotting the trajectory of the system; in other words, the pattern of Puppy's movement changes over time.

We can now bring in another feature of dynamical systems: attractors. Put very simply, these are preferred states in phase space toward which the system will evolve if it's given sufficient time.[3] Attractors come in a variety of forms. So, if Puppy moves with a gait where the values of the joint angles continually repeat themselves over time, say, a steady walking gait, then the system will settle into what is known as a "periodic attractor" (although, as Pfeifer and Bongard point out, given the fact that the joint angles are unlikely to repeat in exactly the same way each time, it is more likely that it will be a "quasi-periodic attractor").[4] If Puppy falls over, and stops moving altogether, so that the joint angles converge on a single unchanging point in phase space, this is a "point attractor" (which isn't really very interesting, as you can imagine). If the trajectory moves around in a particular well-defined region of phase space, but, within the space itself, the exact trajectory is unpredictable, then the attractor is referred to as "chaotic" (to use a slightly looser analogy, this is a bit like someone attempting some very free-form dancing at a party: he may have a particular dancing style that uses particular movements, and all his dances may look similar, but one can't predict exactly which dance will be produced on a given occasion). This is the technical, mathematical definition of chaos, which refers to a state that appears unordered, but which has a precise underlying structure that can be reproduced exactly given the same starting conditions.[5]

Trajectories can converge on a given attractor from a variety of starting conditions, much as a ball rolls to the bottom of a bowl no matter where you place it on the rim. The sum of all the different trajectories that lead to the same attractor is therefore known as its "basin of attraction." Attractor states are interesting because, despite the fact that the system in which they occur is constantly and continuously changing, attractors are easily identifiable as discrete entities. To put this in real terms, although Puppy's joint angles change continuously over time in a smooth fashion, it is nevertheless clear to those of us watching Puppy that it is either walking, running, or standing still, and this would also be clear on our dynamical system graph as Puppy's trajectory moved out of one basin of attraction and fell into another.[6]

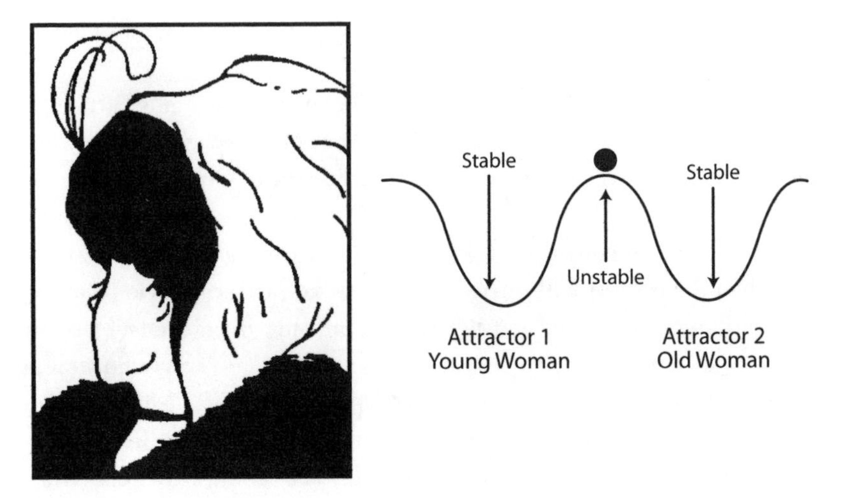

Figure 8.1. Attractor states are low-energy stable states in a dynamical system. To grasp this idea schematically, study the above image. You should see an old woman in profile, but as you continue to look, you should see the image shift to that of a young woman looking away from you. The image of the old woman and that of the young woman are both stable attractor states, and as your perception shifts from one to the other, you are falling out of one attractor state and into the other. You never see both at once, however, because that is not a stable state of the system.

The Sweet Smell of Significance

> Nothing revives the past so completely as a smell that was once associated with it.
>
> —*Vladimir Nabokov*

> I found one remaining box of comics which I had saved. When I opened it up and that smell came pouring out, that old paper smell, I was struck by a rush of memories, a sense of my childhood self that seemed to be contained in there.
>
> —*Michael Chabon*

With this primer on dynamical systems in place, we can move on from the artificial Puppy to Walter Freeman's studies on real rabbits. Freeman investigates the neurophysiology of smell by using simple conditioning experiments that teach rabbits the value of an odor (for example, thirsty rabbits are exposed to an odor and then rewarded with water, so that they learn the association of that particular smell with water). He then monitors activity over the olfactory bulb, the part of the brain that deals with smell, as rabbits are exposed to these conditioned odors.

One of the interesting things that Freeman has discovered is that a rabbit's brain produces a response only to conditioned odors, and not to unconditioned ones. In other words, the smell has to hold some significance or meaning for the rabbit in order for recognition to occur; the smell must offer an affordance to the rabbit. An unconditioned odor goes unrecognized and, one could argue, doesn't count as a smell at all in the environment (umwelt) of the rabbit. So, under more natural conditions, a hungry rabbit that has previously eaten carrots, say, and then encounters the smell of a carrot will generate activity in its olfactory bulb because the smell is part of the affordance of eating that the carrot offers. The smell of something else made of carrots, like carrot and coriander soup, won't trigger olfactory bulb activity because carrots in that form are not meaningful.

As Freeman sees it, findings like these suggest that what a rabbit is doing when it smells is actively seeking to improve its current situation (which links to Dewey's and Gibson's ideas that an organism's behavior is purposeful in this way). The neural connections in its olfactory bulb are strengthened to the extent that what it encounters (e.g., the smell of carrots) serves to satisfy its current needs (e.g., the need for food). Again, this links to a point we made earlier, specifically, that behavior is not about producing the "right" response given a particular stimulus, but often means producing the "response" that subsequently leads to the "right" stimulus.[7] As we've already noted, this reversal of the usual way of thinking about behavior is crucial because it then makes it very easy to decide what the "right" behavior is for a given set of circumstances (i.e., those that improve the animal's current state of affairs). All animals have certain physiological requirements that ensure their continued survival, which have been shaped by natural selection, and this helps to distinguish a "good" perception of the current situation from a "bad" one. So the smell of carrots is good, because it affords eating, but the smell of a fox is bad, because it affords being chased (and possibly eaten yourself). The upshot of all this is that we can expect animals will then act to achieve the right kind of input (the right kind of perception of a situation, in other words), and this perception will then influence what the next desirable input would be (a full stomach may mean that the feeling of sitting quietly in the dark of the warren would be the next desirable input, for example).[8]

Chaos Is the Key

> It turns out that an eerie type of chaos can lurk just behind a facade of order—and yet, deep inside the chaos lurks an even eerier type of order.
> —*Douglas Hofstadter*

Freeman assumes that the initial neuronal strengthening that occurs during the learning process takes place via Hebbian learning (named after the great neuroscientist Donald Hebb, who came up with the idea in the first place),[9] where those neurons that fire together in response to an event become "wired" together and form an association. Alternatively—and following on from our ecological perspective of the previous chapters—we can say that the rabbits are picking up on the carrot odor–water invariant that exists in the experimental situation, so that the rabbit learns of the association between water and odor as a higher-order stimulus.

Using a combination of his empirical studies and mathematical modeling, Freeman has used this insight to come up with the idea that, over the course of learning a particular odor-water pairing, the interconnected neurons in the olfactory bulb form what he calls a "nerve cell assembly" (NCA). Excitation of any part of the NCA by a stimulus tends to cause activation of all of it, and this activity then spreads to include the entire bulb. The reason why this is so useful is that, on any given sniff, an animal receives only a small whiff of the reinforced smell, so that not all receptors will be activated. The strengthening of connections in the NCA to the extent that simulation of any part can trigger the whole means that, in effect, the NCA is amplifying the low-grade signal received from the environment, and this, in turn, provides the crucial mechanism that allows a characteristic odor-specific pattern of activity to spread over the bulb.

Modeling studies have shown that the formation of the NCA, along with the excitation of the bulbar neurons, primes the entire olfactory bulb to form a chaotic attractor that corresponds to a particular odor-reward pairing. A new attractor is formed for each new significant odor-reward pairing so that, as a result of its experiences, the rabbit's olfactory bulb takes on a specific configuration—an "energy landscape"—that consists of several basins of attraction, each one corresponding to a specific class of learned combinations of odor and reward.[10] Interestingly, as new attractors are formed, all the other attractors rearrange themselves over the bulb, so that there is no single fixed pattern of attractors across time: the landscape of bulb activity is constantly shifting. Each new significant experience therefore changes, to some degree, the significance of all the rabbit's previous experiences. This, in turn, influences the nature of the rabbit's future odor experiences. In other words, we can see the mutually determining nature of the system: new experiences lead to changes in the rabbit's brain, which then change its experience of its environment, and how it then acts, which changes its brain, which changes its experience of the environment. How an animal responds to the world (by forming a new attractor for a new significant odor-reward pair) influences how it subsequently perceives the world just as the "loopy" mutualistic theory of behavior we have discussed previously suggests that it should.

Most interesting of all, from our point of view, is Freeman's suggestion that the chaotic attractors alone are the patterns of neural activity that get sent from the bulb to the rabbit's cortex. This is significant because attractors are highly individual patterns of activity that reflect the sum total of the rabbit's experiences with a particular stimulus, the contexts in which these occurred, and the significance they hold for the rabbit, and they also bear the signature of the rabbit's experience with other odor-reward pairings (because of the overall changes induced in the attractor landscape as new odors are learned).

The one thing that doesn't get sent forward to the cortex is the pattern of activity that was initially generated by the carrot odor molecules themselves as they contacted the odor receptors of the nose and triggered a neural response. By the time the chaotic attractor has formed, all trace of any odor-receptor stimulation has disappeared completely. If Freeman is right about this, then it has immense significance for anyone trying to link brain processes to psychological-cognitive processes: the olfactory bulb attractors cannot be said to "represent" either the odor of a particular object, say, a carrot, or any other aspect of a carrot per se, because attractors incorporate more than just carrots; they also include the context in which the smell of carrots became meaningful to the rabbit.[11] The patterns produced across the bulb correspond to the significance that a carrot currently holds for a rabbit, based on its past history of experience with carrots, and not to the carrot's essential "carrotyness" as an object.

Moreover, it should be clear that the formation of the NCA and the attractor states in the olfactory bulb show exactly the kinds of rich temporality that Wheeler suggests are important in noncomputational, dynamical systems. The rate at which the NCA forms, for example, will influence the timing of attractor formation, and formation of the NCA itself is dependent on the temporal characteristics of the rabbit's rhythmic sniffing, because this determines the rate at which odor molecules are able to contact receptors. There is therefore a direct and dynamic coupling linking the rabbit's body (its sniffing nose), its brain (which is falling into attractor states with each sniff), and its environment (the affordances offered by the object that is being sniffed).

We can use Gibson's metaphor of "resonance" to characterize this process. That is, the rabbit's bulb becomes selectively "tuned" to "resonate" to the invariants that are significant for the rabbit.[12] As resonating systems respond differently to the same stimuli depending on how they are tuned, this picks up on the changes in the attractor landscape as new significant odors are encountered, and it can also explain why different animals pick up on different invariants in the world (with the caveat, of course, that the notion of "resonance" is not to be taken too literally). It is this combination of features—that the brain and nervous system "resonate" to the world, and that activity patterns are responsive to the significance of an odor, rather than to the odor itself—that couples perception and action together.[13] Freeman's argument is that the motor system is guided by the global states of activation in the olfactory bulb because they are all linked together as a loop: a carrot that affords eating is priming the motor system to produce the appropriate response as part and parcel of the rabbit's perception of this affordance. This emphasizes that how animals act is at least as important as how they think. Indeed, in this approach, acting simply is a form of thinking, and we don't need to posit any kind of linear internal, representational input-processing-output procedures to account for the organized patterns of behavior that organisms display. Andrew Pickering, a sociologist of science, who has written a wonderful book about the British "cyberneticians," including Grey Walter, suggests that we should refer to the brain and what it does as "performative" rather than "representational," which captures exactly this point.[14]

A mechanism like this also potentially enables an animal to achieve certain goal states, without ever having to "represent" them in advance or even to have any sense of the goal state it is trying to reach.[15] Rather, the animal may simply feel a "tension" in its sensorimotor system as it moves away from an equilibrium point (as activation is caused to move out of one basin of attraction and closer to another, for example), and is drawn to movements and actions that lower this tension, without any overt recognition that this is what it's doing, or why. This clearly works for humans on occasion. One example I can give of this comes from when I used to live in Liverpool. Every morning, I would go for a run from my house, around Calderstones Park and back again. On the way back, I would run along one particular street, and, although I entered the street on the pavement, I would inevitably find myself switching to running in the road itself about halfway along. Only very gradually did I become consciously aware that I was doing this, and, even once I realized what was happening, I still didn't understand why it was that, along this particular stretch of this particular street, I ran along the road (and not the far safer pavement). Then one day I saw it. About halfway along, as the street rounded a small green, the pavement became increasingly angled. Running along this angled pavement threw me out of my rhythm—in Dreyfus's terms, you could say it was creating a tension in my sensorimotor system and pushing me away from the nice stable attractor state that was my running gait—and so moving onto the flatter road surface allowed me to improve my current state of affairs with respect to running comfortably and efficiently, even though for most of the time I had been doing this, I had been aware neither that this was what I was doing nor why. Even once I knew what I was doing and why, it was often the case that I would end up switching to the road with no anticipation that I was going to do so, nor any recollection of the moment that I made the decision.[16]

Things Are Not Always as They Seem

> I could be bounded in a nutshell, and count myself a king of infinite space.
> —*William Shakespeare*, Hamlet

> Logic will get you from A to B. Imagination will take you everywhere.
> —*Albert Einstein*

So a dynamical systems approach shows how we can investigate and explain intelligent action in the world without being forced into a representational-computational straitjacket, and it allows us to bring the body and the world into the cognitive system in a natural and useful manner. This complements the ecological psychology approach we highlighted in chapter 6, where we considered how the structure of the environment itself could provide useful information to an organism in highly cost-effective ways.

One thing to make clear, however, is that, despite our critical assessment of representational theories, it would be wrong to leap to the conclusion that representations don't exist at all and/or are not useful in accounting for behavior and action (even if we don't yet know precisely what they are). As Andy Clark has suggested, there are many aspects of the human world (at least), that are "representation hungry."[17] Clark uses the example of asking someone whether they think it is ethically acceptable for the United States to manufacture more guns than it can sell legally. This cannot be dealt with through the use of any form of sensorimotor, embodied process alone; it necessarily requires symbolic representations. One can't deal with issues like morality, the arms trade, crime, and business practices without reference to what John Searle (another philosopher of mind) has called "institutional facts": things that don't exist in any concrete, physical form but about which we all agree, and which we use to structure our lives (money, for example, is an institutional fact: we all agree that a certain piece of paper counts as currency in a certain country, and can be used in exchange for goods and services even though the paper itself has no inherent value).

Similarly, the epigraphs at the beginning of this section concern a "representation-hungry" process: imagination obviously requires the ability to conjure up aspects of the environment that aren't present or may not even exist. So this is not necessarily an argument for doing away with representational thought altogether. The alternative viewpoint expressed here just shows how this remains an open issue, one for which we don't yet have a good answer.[18] Specifically, it is a strong argument for a pluralistic approach; especially if we wish to provide ourselves with a means for studying other animals besides ourselves. A pluralist approach means recognizing that some kinds of activities may require representations and maybe some won't, and also that some kinds of activities will require both a nonrepresentational component and a representational one as part of exactly the same process. Perhaps it is simply not an either/or argument.

Even with this caveat in place, and acknowledging that representations may be neither irrelevant nor nonexistent, we can still advise some caution when taking the representational view, because it often forces a distinction between organism and environment that, in all likelihood, simply doesn't exist for the organism itself. Even if we can identify certain internal processes in the brain that are correlated strongly with external behavior, it would be wrong to conclude that these constitute representations of the world that are used to guide the animal's behavior. It is wrong because we are using our external frame of reference to view both the internal mechanism and the external behavior simultaneously, and it is easy therefore to map the one directly onto the other. For the animal concerned, however—the animal that is actively engaged in behaving—the internal activity that we see from the outside simply is that animal's experience of the world, not its representation of it.

A Frog's Eye View

This needs a little unpacking. To do so, we can draw on some classic experiments in neurobiology conducted by the Nobel Prize winner Roger Sperry.[19] In these experiments, Sperry took a tadpole and sectioned its optic nerve so that he could rotate one of its eyes by 180°.[20] When this tadpole grew up into a frog and attempted to catch a fly using only its upside-down eye (with the good eye covered), it was unable to do so. If the fly appeared in front of the frog, the frog's tongue would fly out behind it;

if the fly appeared below the frog, the frog's tongue would shoot out above it. In other words, the frog behaved as though its eye was actually the right way up. So, despite their new position in the upside-down eye, the nerve cell axons that linked the eye to the relevant part of the brain (the tectum) had returned to their original position in the brain as they regenerated. No matter how many times it failed to catch a fly, the frog simply could not adapt its behavior and move its tongue any differently. Consequently, as far as the frog is concerned, "there is no such thing as up or down, front and back, in reference to an outside world, as it exists for the observer doing the study."[21] Instead, there is only an internal sensorimotor correlation between the position of the image on the frog's retinal map and the movement of the tongue. The frog has no notion of the external world as it appears to us, and no sense that its eye is upside-down as we see it.[22]

Now, notice we used the term "retinal map" above; this refers to the fact that nervous systems show what's known as "topological mapping" (the London Underground [subway] map is a topological map: a simplified diagram that preserves all the essential details about the relation of different stations to each other). In essence, this means that neurons that are next to each other will fire in sequence when a stimulus moves across adjacent positions in the sensory field, so that one can often recognize what the organism is viewing from the pattern of neural activation produced. But if organisms form neural "maps" of this kind, then surely these are "representations" of the world, just as the London Tube map is a representation of the real underground train system? Well, no. If we look at neural somatosensory or sensorimotor maps in that way, we are making precisely the mistake outlined above. We can see these mappings because we are able to look at both the inside of the frog and the outside world at the same time, and to see how they correlate with each other. From the "inside"—from the perspective of the frog itself (and indeed for us humans in our own acts of perceiving)—these mappings are not representations of the world; they are simply that animal's experience of the world: they are what the world feels like and how it "shows up" for them.[23] There is no extra "layer" of representation (after all, and making the same point we quoted from Gibson earlier, if there were maps of the world like the Tube map, who or what inside the frog's brain is looking at them?).

All this talk of internal mappings may seem to contradict the ecological psychology approach we described in chapter 6. After all, we spoke there

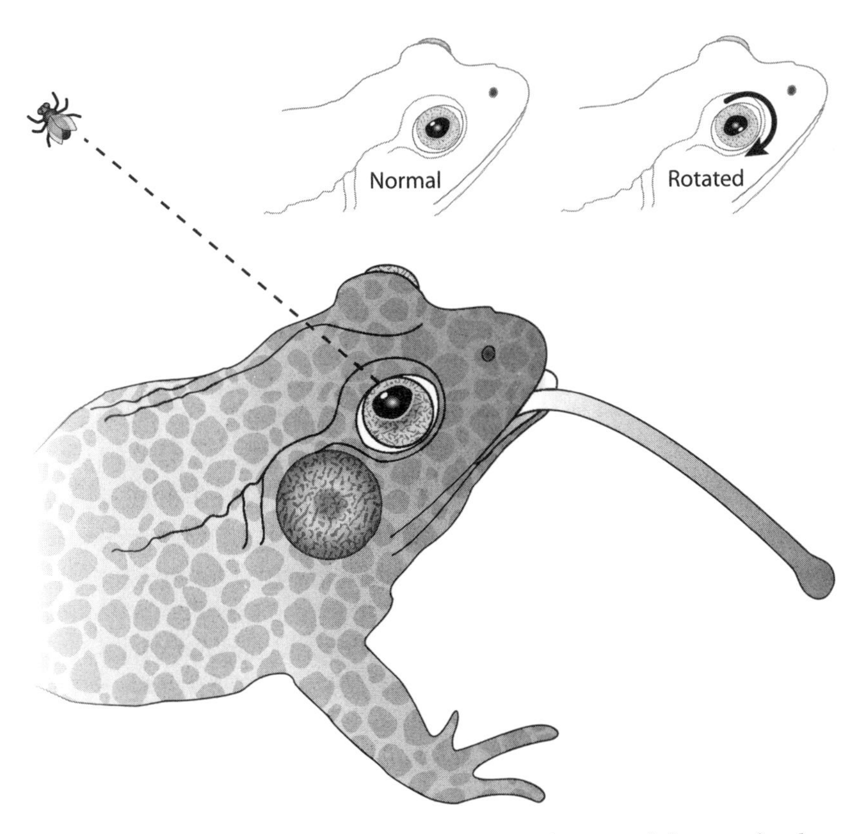

Figure 8.2. In Roger Sperry's classic experiments, he rotated the eye of embryonic frogs by 180° and then allowed development to proceed. When the frogs were grown and attempted to catch flies, they would stick their tongue out in exactly the opposite direction from that in which the fly was located.

of organisms' being able to contact and perceive the world "directly." How can this be reconciled with the actions of the upside-down-eye frog that apparently has access only to its own internal mappings? The key things to remember here are, first, that the actions of the frog are always directly guided by the world (the presence of prey items), and, second, as we've said before, an ecological approach doesn't deny the existence of an internal nervous system that contributes to cognitive processes. The ways in which environmental stimuli are encountered will depend crucially on the structure of the nervous system and how it has been shaped over evolutionary time (not least by the kinds of information available in the environment). The frog's umwelt, for example, includes small, dark moving objects, and its perceptual system can pick up the invariants of a flying insect that afford catching with its tongue. Its umwelt doesn't, however, include other features of the spatial layout, and so its perceptual system is not tuned to pick up the invariant structures that provide information about these, which would allow it to adjust its behavior in a way that would allow effective fly catching.

Put more generally, and to echo the point made in the previous chapter, the argument here is that nervous systems are truly integral to the world as it is experienced by an organism, and are not merely the means by which the world "out there" is detected. To say that the frog has access "only" to its internal mappings is to make the same mistake we've been talking about this whole time: making a false distinction between the organism and the environment. We can make this distinction because we exist outside the frame of reference of the animals we study, and it explains why we're so successful at describing the behavior of other animals scientifically (more on which below). It's often said, for example, that fish "don't see the water" in which they swim, and Ludwig Wittgenstein once pointed out that one can't look through one's spectacles and at them at the same time. Similarly, nervous systems and bodies are an inclusive part of the world to which an organism has access, but no organism can "see" this for itself. Instead, it just experiences the world in a particular kind of way. John Haugeland, an eminent philosopher/cognitive scientist, refers to this as the "intimacy" that exists between organism and environment.[24]

Being in the World

> We do not say: Being is, time is, but rather: there is Being and there is time.
> —*Martin Heidegger*

> I never lose sight of the fact that just being is fun.
> —*Katharine Hepburn*

We are not, however, used to thinking of ourselves and other animals in this way. We do not notice the intimate, and inextricable, twining of our bodies with the world because, ironically, when we're thinking about our place in the world, we have to stand back from it; otherwise we couldn't contemplate ourselves in this way. This detached way of looking at things, in the Western world at least, can be traced to the philosopher René Descartes, author of the famous dictum "I think, therefore I am," who suggested that the only thing we could really be sure about was the existence of our own thoughts (specifically, while all our senses could be deceived in various ways, and the outside world and our own existence could thus be merely an illusion, the fact that we can entertain thoughts that cast doubt on the very existence of our thoughts must prove that, logically, we exist because there has to be something there that is having that very thought). This in turn led to the position that the mind is a completely different kind of entity from the physical matter that makes up our bodies. Nowadays, "substance dualism," as this position is called, has fallen by the wayside in favor of materialism (basically, the position that the mind is the brain, and therefore of essentially the same physical stuff as the body), but we still retain the idea that we possess "minds" with which we look out at a world that is completely external to us and that we can access only indirectly via our representations of it. This is itself a form of dualism because it makes a clear demarcation between an organism and the world in which it lives, and, as we've seen, this form of dualism is just as misleading as substance dualism.

An alternative viewpoint is that expressed by a more recent philosopher, Martin Heidegger,[25] who argued that, fundamentally, we are not set apart from the world (as "subjects directed to a world of objects"), but we are always in the world, right there in the thick of it, and there is simply no "outside" view.[26]

More specifically, Heidegger's view is that our usual, everyday means of encountering the world is a nonrepresentational form of "smooth coping." In his most famous example, he describes how, when we use a hammer, we do not view ourselves as a subjective mind directed toward the hammer as an object; to hammer effectively, we do not need to represent to ourselves, "Here I am, hammering away with a hammer." Rather, we are simply hammering. The hammer becomes "transparent" to us if all is going well, and so we don't notice any separation between ourselves and the world at all: the hammer is, in Heidegger's terminology, "ready-to-hand." Perhaps a more applicable example today is the use of our computer keyboard, mouse, and joystick. When we are absorbed in our work, Web surfing, or playing video games, we are not aware of the mouse, keyboard, and joystick as objects: we are in direct contact with the environment of the game, not with our means of controlling what appears on the screen. According to Heidegger, most of our everyday cognition takes this form.

If the head should suddenly fly off the hammer—or the joystick stops responding—then we do become aware of the object and our own view of it: the hammer becomes "un-ready-to-hand" (these constructions may seem very awkward, and they are, but Heidegger was trying to get across his concepts in a way that didn't lead straight back into a Cartesian way of looking at the world, because our standard language inevitably leads us in this direction). This is also something we encounter in our everyday lives, because we have to deal with this disruption to our smooth coping when things have broken, have gone missing, or simply are in the way.

Heidegger also identified a third way in which humans view and act in the world, which he described as the "present-at-hand." To do science properly, for example, one has to take a certain stance toward objects in the world, a stance that is completely different from the way we deal with many of the same objects in our everyday practical dealings with the world. In this mode, we become the kind of full-blown "representing" subject with whom Descartes was concerned: an individual who wants to gain knowledge of, and to predict and explain, an apparently external, independent world, and who accesses this world by forming representations of it.[27] Descartes wasn't wrong about our subjective status; his mistake was to assume that the "present-at-hand" is our "default" state, when, as Heidegger argues, it seems more likely that this is a highly specialized trick that we humans can achieve some of the time, under specific circumstances.

This notion of the "present-at-hand" is why the issue of animal minds becomes problematic for us, I think. We assume that other animals must show the precursors of the kind of "present-at-hand" thinking that we use when we are trying to conceptualize, predict, and explain the nature of their cognition, and we do so because we mistakenly assume that this kind of thought is characteristic of all human cognition. We assume it represents our default state for dealing with the world, and so we marry this to the assumption of evolutionary continuity. But our ability to take this specialized, detached stance may well be a recent human innovation, the product of biology and culture (aided and abetted in particular by complex spoken language)[28] coevolving together over the course of our specifically human history, with no precedent in the animal kingdom as such. Instead, the evolutionary continuity between ourselves and other organisms is more likely to lie in the kinds of smooth coping that make up most of our everyday experience, not the specialized, highly reflexive forms of thought that we use to think about our own thoughts and the possible thoughts of other animals.

Similarly, the real problem with a heavily computational view of the brain-mind, with its necessary reliance on representations, is that it also takes the "present-at-hand" as the natural state of affairs, and places cognitive processes at a remove from the animal's brain and the rest of its nervous system, its body, and the world in which it lives. Trying to use the this particular view to decipher what kinds of minds other animals might have just ends up heaping error on top of error. We are doomed never to gain a satisfactory insight into how other animals act and understand the world if we persist with such a view, because intelligent, flexible behavior is never exclusively the product of cognitive processes occurring in the brain. There is no such thing as a naked brain.

CHAPTER 9

World in Action

> The meaningful objects . . . among which we live are not a model of the world stored in our mind or brain; they are the world itself.
> —*Hubert Dreyfus (1972)*

Our brief (and somewhat philosophical) excursion into the brain and its metaphors has brought us to the following position. We have identified some problems with a particular "classical" view of cognition as being purely the rule-based manipulation of symbolic representations, because it tends to underestimate two factors that, as we have seen, appear crucial to an understanding of how animals engage with their environments in flexible ways. These factors are (1) the nature of an animal's body—quite literally, how it is physically put together—and (2) the ways in which a particular kind of body affords certain kinds of interactions with the environment, while simultaneously constraining the performance of others. Taking on board the arguments of people like Jacob von Uexküll and J. J. Gibson, we have, therefore, embraced what we could call a more "relational" and "embedded" view of cognition, where organism and environment are looped together and one cannot sensibly be viewed in isolation from the other.

Bringing Back the Body

> You'll never plough a field by turning it over in your mind.
> —*Irish Proverb*

Action-Oriented Representation

Another way of looking at this is to make the distinction offered by Andy Clark.[1] He identifies the "classical" view as one that understands the brain as a "mirror" of the environment, whereas the "embodied" perspective we have been pursuing sees the brain as a "controller" of action in the environment. In the first case, the separation between perception and action is made very clear: the brain stores "passive" inner descriptions (representations) of the external world that are then manipulated and processed to produce an output that is fed to the motor system, which can then produce action in the world. In the latter case, this distinction between perception and action disappears because an animal's inner states are not passive "pictures" of an external world, but are, instead, plans of action for engaging with the environment.[2] In this way, the distinction between an "inner subjective world" and an "outer external world" that it maps onto disappears also. Clark refers to these plans for engaging with the environment as "action-oriented representations," and we will stick with this terminology for the sake of clarity, despite the problems we discussed in the previous chapter; we just need to remember that from the animals' point of view there is no inside-outside distinction between its action-oriented representations and its experience of the environment.

We can return once again to the world of robots for a very elegant example of action-oriented representation. Roboticist Maja Mataric built a robot rat with the ability to construct an internal "map" of its surroundings so that it could find its way around a cluttered environment.[3] The map was not, however, the kind of map with which we're all familiar: one that depicts the layout of a landscape in some kind of "pictorial" fashion. Instead, the map consisted of a combination of the robot's own motion and its sensory readings as it moved around the environment. When the robot came up to a wall, this wasn't represented in the map as, say, a "solid vertical object." Instead, it was stored as a combination of what the rat was doing at the time the wall was encountered and what it could sense—e.g., "moving straight, with short lateral distance readings heading south"—so from our outside observer-oriented perspective this can be seen as a map of the layout of the environment, but, from the rat's point of view, it is simply an action plan (again making our point about the world as perceived by the organism versus outside assessments).[4] As the robot moves around, the "nodes" that correspond to particular landmarks become activated as they are encountered, and activation spreads in the direction that the robot moves, so that the robot generates an "expectation" about the next landmark it will encounter (there is no suggestion that this is conscious in any way, of course). If the robot then "wants" to reach a particular location, the "node" that corresponds to this location is activated, in addition to the rat's current location. The spreading activation through the map produced by the nodes naturally will result in the shortest route between them being found. As a result, there is no work for "cognition" to do: the perceptual map doesn't need to be transformed into an action plan by any cognitive process because the locations in the map are already specified in terms of the movements needed to reach them, along with the correlated perceptual input to the robot (what it sees as it moves around). In the words of Andy Clark, then, "the map is its own user."[5] This action-oriented map illustrates perfectly how perception and action can be so thoroughly entwined that there is simply no room left for "cognition." What is most fascinating is that, if one takes a real rat and carries it around a novel environment, while preventing it from moving its legs, there is no change in the activity of its hippocampus (the part of the brain associated with spatial mapping of this kind), suggesting that motion is crucial to the generation of a map in real rats as well as in robots, permitting us to hypothesize that the former may use similar kinds of "action-oriented" representations.[6]

Andy Clark gives some neat human examples that also suggest an action-oriented perspective is warranted. One particularly compelling study asked participants to wear distorting lenses while throwing darts at a dartboard.[7] The lenses had the effect of shifting the visual scene slightly to one side so the participants tended to throw wide of the mark initially, but, with repeated practice, the subjects were able to adapt to the side-shifted view and could hit the board quite accurately. However, they could do so only if they used the same arm and the same overarm throwing action. They couldn't use their other arm at all to throw darts that would hit the board, nor could they hit the board if they threw the darts underarm. The adaptation to the lenses was utterly specific to the particular perception-action loop associated with the use of the dominant-hand, overarm throwing that they had trained on.

These findings do not sit well with the standard view of separate, independent perceptual and action systems. If that view were accurate, then, given that only perception was distorted, the participants who had adapted to their new perceptual inputs should have been able to throw in all kinds of different ways because the adapted perceptual system would now be sending the right kinds of signals to the motor system, which—being independent of perception—would be able to respond to those signals appropriately. The fact that participants showed adjustment to the lenses only when they threw the dart in a particular way suggests not that perception and action are two independent systems, but that both work together as a tightly coordinated, fully integrated unit; the adjustment was made to the particular perception-action loop of overarm dart throwing; hence there was no general pattern of adjustment. This also fits with Gibson's ecological approach to psychology, which takes as its starting assumption that animals' perceptual systems are geared for action in the world—foraging, finding shelter, avoiding predators and the like—and not to creating a replica of the world inside their heads.

Research in my own department also points to the value of this action-oriented point of view.[8] Specifically, there is an intriguing difference between expert musicians and nonmusicians in their susceptibility to what is known as the "missing fundamental illusion." John Granzow, who studied this phenomenon for his master's degree, is an expert musician himself, and the idea that musicians might actually hear the world differently from nonmusicians seemed worth probing in more detail. To oversimplify hugely—and no doubt to John's horror—the "missing fundamental illusion" refers to the way in which our perception of pitch, or the fundamental frequency of a tone (whether we hear it as low or high), does not depend on the presence of sound energy at that frequency as long as we are also supplied with the set of overtones (or harmonics) that correspond to that fundamental frequency (overtones are what give a tone its timbre—the quality of sound that lets us know that we're listening to a trumpet and not an oboe, for example). In other words, we can hear pitch, even when we're presented only with timbral information.[9]

In the study that got John interested (and again to oversimplify), human subjects were presented with pairs of tones in which pitch and timbre had been manipulated in such a way that the experimenters could identify whether people were able to track the missing fundamental or were instead responding to the overtones.[10] People were asked simply whether the second tone in a two-tone sequence was higher or lower than the first. The tones were manipulated so that the overtones would indicate the tones were rising, while the fundamental frequencies indicated the tones were falling (and vice versa). By placing them in opposition in this way, the experimenters could identify whether people were making judgments of pitch based on the fundamental frequency or using the overtones to do so. The results showed that expert musicians were very good at tracking the missing fundamental, and so got the answer "right" (i.e., they identified the tones as rising or falling based on the fundamental frequency), whereas nonmusicians tended to track the overtones, so mistaking a difference in timbre for the difference in pitch (i.e., they would identify the tones as rising on the basis of the overtones, when the fundamental frequency of the tones actually fell in the sequence). (I was especially bad at this test when I took part in John's replication of the original work; I just about scraped a D . . .)

When John, aided and abetted by his supervisor, John Vokey, replicated and expanded on this study, he found something unusual. The expertise effect could be eradicated completely when people were placed under a time constraint. Only when people were allowed as long as they wanted to respond did musicians score better than nonmusicians. A chance remark by a nonmusician who nevertheless did quite well on the test helped to kick off a new train of thought. The nonmusician said that he'd hummed the notes to himself after hearing them, and in this way he could tell whether the change went up or down. Indeed, the humming often reversed his initial impression and even altered his memory of the tones. Reflecting on this, the two Johns realized that perhaps what allowed musicians to succeed wasn't a difference in hearing but rather the kinds of experience they'd had actually playing and performing music—actions might be as important as sensory input. John G., when he tunes his guitar, for example, often hums both the reference tone he wants to achieve (as produced on a piano, say) and the tone from the string he wants to tune, as a way of knowing whether he needs to go up or down. So, when musicians can hum to themselves—using their voices as another kind of "instrument" to help them disambiguate the tones—they get the answer right, but when speeded-up answering means they don't have time to hum the tones back to themselves, the advantage disappears. A further experiment to test this found that this was the case; the expertise effect was restored when musicians were not only allowed more time but specifically encouraged to hum. Indeed, not only was accuracy restored, but the musicians produced near perfect results.

Again, these findings don't fit well with conventional unidirectional models of hearing, where a sound goes in, a response comes out, and nothing about the response can change the nature of what a person perceives. Instead, they correspond much better to the ecological, "loopy" feedback theories we discussed in previous chapters, where a person's response will inform her subsequent perceptions, and so by their own actions people can genuinely change what it is that they see or hear. As John G. says, however, most auditory experiments simply assume that only a unidirectional relationship exists, and so are specifically designed in a way that doesn't allow for the identification of more embodied perceptual feats. We could be missing out on many interesting action-oriented phenomena owing to these kinds of systematic biases in our experimental designs.

Fast, Cheap, and out of Control

Viewing representations as some kind of action plan, rather than as passive pictures of the world, fits nicely with our exploration of ecological psychology, with its emphasis on affordances, and on the dynamical systems approach, where internal and external resources mutually enable and constrain each other. If we think back to earlier chapters, we can also see how the sensors and motors of Elsie and Elmer, the cricket robot, and the didabots can be understood as the forerunners of the more elaborate action-oriented representations we've described here. Moreover, these examples made it clear that the internal modeling of the world is often completely unnecessary for the production of interesting and complex behavior.

This approach has been pursued even more vigorously by Rodney Brooks,[11] a roboticist at MIT, whose motto for a long time was that his robots should be "fast, cheap, and out of control." Brooks first became interested in robots as a schoolboy in his native Australia; he got hold of a copy of Grey Walter's book and built his own tortoise, using transistors, rather than vacuum tubes. His robot, Norman, could wander around the floor, respond to light, and shove its way past obstacles, just like Grey Walter's machines. Later on, at Stanford University, Brooks became involved in more "traditional" AI-style robotics that used the classic "sense-represent-plan-act" approach, in which the robots were equipped with internal representations of the environment, and computed solutions to a task before executing them. The problem with such robots is that they were very slow as a consequence. The robot that Brooks worked on, for example, moved so slowly that the movement of the sun across the sky, and the changes in the shadows thrown, confused its internal representations![12] Brooks found this very disappointing and frustrating. Grey Walter's robots had been made on the cheap and were capable of all kinds of interesting behavior in a changing world; at Stanford, there were massively expensive robots that, internally, were capable of so much more than a tortoise, yet they didn't operate nearly so well. As he notes, "to an external observer, all that internal cogitation was hardly worth it."[13]

Consequently, when Brooks began his own lab, he returned to the original inspiration of Grey Walter's tortoises. Brooks's goal was to design robots that were capable of coping with changes in a dynamic environment in a robust and efficient way, without having to build in any kind of central processor (i.e., a brain), so dispensing with the need for large amounts of more costly, internal electronic circuitry, and eliminating the slow process of computation.[14] More specifically, Brooks decided he would attempt to produce this kind of "insectlike" intelligence in his robots, on the grounds that most of evolutionary history has been spent refining the perception and action mechanisms that help animals move around and engage actively with the world.[15] "Higher" cognitive faculties like language, problem-solving behavior, reasoning, and expert knowledge all appear relatively late in the day, evolutionarily speaking. This, in turn, implies that all these higher faculties—those that we consider to be the most complex—must actually be quite simple to implement once the essential perceptual and motor processes that enable an organism to act in the world are available. Not only that, but these perceptual and motor processes must, as a direct consequence, underpin the evolution and elaboration of the "higher" functions themselves, so they are not free of bodily influence in the manner we tend to assume, but will also be "action-oriented" in the way we discussed above. Our understanding of "higher" cognition, in other words, has to be grounded in a firm understanding of the ways in which perception and action mechanisms lead to adaptive behavior in dynamic environments.

Brooks's robots use what is called a "subsumption" architecture, in which a complex behavior is broken down into a series of simpler behavioral "modules" or systems, each geared to meeting a specific goal. Each system "sees" the world in a way that is entirely different from how any of the other systems "see" the world. These modules are then organized into "layers," where each layer's goal is "subsumed" by the one above it (we have already met something like this in the form of Elsie's and Elmer's light and touch sensors: if you recall, when the touch sensor was activated, the signal from the light sensor was effectively "ignored").[16] Importantly, there is no need for any of the systems to be integrated to form any kind of coherent concept of the world. Using this approach, Brooks came up with, among others, Herbert (named after Herbert Simon), a robot who could collect empty Cola-Cola cans and carry them away. For Herbert, perceiving an obstacle with its ultrasonic detectors was simultaneously an act of stopping movement. Herbert didn't "see" objects as such, nor did it have an internal representation of an obstacle as a solid object in the real world. For Herbert, an "obstacle" was simply the specific action of stopping dead. Using exactly these principles, Brooks went on to develop the Roomba, the highly successful vacuuming robot, which can detect dirt, avoid obstacles, and, again like Elsie and Elmer (but much more effectively), recharge itself when its batteries run low.

With all of his robots, Brooks's goal was to illustrate that one didn't need "cognitive processing" or any form of internal symbolic representation in order for the robots to function effectively. The close coordination of perception and action, and the coupling of these perception-action loops to the environment, meant there was simply nothing left over for "cognition" to do. Instead, his robots "used the world as its own best model."[17] The behavior of the *Portia* spiders we discussed in chapter 4 can be seen in this light; their patterns of scanning and movement, combined with their limited neural capacity, suggest not that they form "pictures" that "mirror" the world, but instead that they may use an architecture similar to that of Brooks's robots (in terms of function, if not in terms of how it is built). The interdigitated patterns of scanning and movement seen when the spiders were given a more realistic setup seem to point to the idea that they may also use the world as its own best model, and that their patterns of starting and stopping are dictated by a similar close coupling between perception and action. As we saw, perceiving a gap in a horizontal route was simultaneously an act of changing direction for the spider because of the servomotor mechanism linking its eye movements to the degree of body movement. As with Brooks's robots, this led us to the conclusion that there was actually no need to posit an internal cognitive process of "planning": the spiders could successively approximate a route toward prey as they scanned, turned, and traveled, scanned, turned, and traveled.

Parallel, Loosely Coupled Processes

A more general way of referring to the kinds of subsumption architecture used by Brooks's robots is that they involve "parallel, loosely coupled processes."[18] Rather than constituting a sequence of independent processes following one after the other (i.e., the robot senses the world, represents it internally, plans what it's going to do, and then acts), the robot's different modules and layers are coordinated with each other only "loosely," that is, via the environment, and not by any kind of internal computational process. But what do we mean by this? How can an animal's behavior be coordinated externally by the environment, rather than internally by the animal? Of course, when we ask this question, we are already falling back into the trap of thinking that the organism and the environment are two completely separate entities, rather than mutually entwined and so "codefined." That's why it seems so odd, but if we accept that the skin (or exoskeleton, as we'll see in this case) isn't all that important a boundary, we will begin to find it much more natural to think in terms of the mutuality between animal and environment.

Keeping in Time

It has long been known that, in creatures like stick insects, each leg is independently and autonomously controlled, and there is apparently no center in the insect brain that coordinates their movements. If there is no way for the brain to "know" what any particular leg is doing, however, nor for the legs themselves to have any "sense" of the movements of the other legs, how do insects ever manage to produce any form of coordinated leg movement, rather than simply falling over or failing to move anywhere at all? This is where "loose coupling" comes in. The legs achieve coordination by exploiting the way that they all interact with the environment.[19] When one leg is moved backward, for example, so generating the power needed to move the body forward, it causes all the angles of the joints of the other legs to change automatically and simultaneously (remember our example of the Puppy robot). As the body then moves forward, all the legs are automatically pulled forward as well, and, again, the joint angles adjust accordingly.

The inevitability of these changes in joint angle can thus be exploited by the insect, and coordination between the legs immediately becomes possible: global communication is achieved because all the legs interact with the environment at one and the same time. Neural connections inside the insect aren't needed at all. It is both cheaper and more efficient to exploit the structure of the environment and its effect on joint angles than to build a more complex neural network into the insect itself that can monitor its leg movements. Similarly, we can argue that the neurons in the rabbit's olfactory bulb are loosely coupled through the environment: the neurons "cooperate" with each other and so settle into a particular attractor state via the presence of a particular significant smell in the rabbit's environment.

Loose coupling through the environment can also help reduce the need for any specific cognitive control of behaviors that occur between animals. Take rat pups, for example. Early in life, rats pups are pink and hairless and can't regulate their body temperature effectively. When their mother leaves them, a litter will huddle together in various configurations that help to compensate for the drop in body temperature. Jeff Alberts, a neurobiologist, who has spent many years studying these behaviors in detail, arrived at the conclusion that the complexity of the pups' behavior was underpinned by some very simple rules. For rats up to 7 days of age, the rules seemed to be "Stay in contact with a vertical surface"[20] (resulting in wall following or "thigmotaxis") and "Move toward warm objects." For those between 7 and 10 days, a third rule seemed to come online: "Do what your littermates are doing" (i.e., if they are active, be active; if they are inactive, be inactive).[21] Using computer simulations, Alberts and his postdoc, Jeff Schank, found that, when they moved their simulated rats according to the first two rules, they behaved just like 7-day-old rats, and when they added in the third rule, they could get the stick images to behave like 10-day-old rats.[22] The coordinated behavior of rat pups younger than 10 days comes about via their interaction with the environment; each rat pup follows these same "rules," and, as a consequence, they end up huddling together without even having to possess any sense at all that there are other rat pups present.

These ideas relating to loose coupling are brought out even more strongly in work using "roborats," rather than just computer simulations. These are very simple approximations of rat pups that look like slightly pointy toy cars (with the pointy end at the front, just as in a real rat). What is most interesting from our point of view is that, in one experiment, the rat robots were given a completely random control architecture—that is, they had no rules built into them at all.[23] Nevertheless, when they were placed in an arena and let loose, the patterns of huddling they showed were either intermediate between or identical to those shown by the 7- and 10-day-old rat pups, and looked just like those produced in the simulations where the "rules" were being followed.[24] As the researchers looked more closely at how the robots behaved, it became clear that the roborats' behavior was being coordinated by the way they interacted with the environment. When the robot contacted a wall, its tapering nose caused it to slide along it, with the direction determined by the angle at which it had contacted the wall. The options for other kinds of movement (that would get it away from the wall) were constrained by this contact, resulting in wall-following behavior. If the robot encountered a corner of the arena, its ability to move became even more limited: basically its only option was to move backward. Even this option was prevented, however, if other robot rats randomly encountered the robot while it was in this position. As the other robots pressed in from the sides, the classic "huddling" and "corner-burrowing" behaviors shown by real rat pups then emerged, even though there was no sense in which any of the rats had the "goal" of huddling. The coordination was achieved through exploitation of the environment.

Of course, it is extremely important to remember that this does not mean that real rat pups have only a random architecture, nor were the researchers conducting the study trying to demonstrate that this was the case. What it does show, however, is that rat pups need not be equipped with any dedicated sensorimotor routine or specific neural processor that produces their characteristic behaviors: they don't need to be "programmed" with specific "rules." Rather, the loose coupling achieved via the environment is sufficient to produce realistic behavior. Moreover, and this is crucial point, the loose coupling was dependent on the structure of their bodies—the environment alone can't achieve the coupling; the animal itself has to be "set up" in the right way as well. The robots' pointy ends were essential to producing wall following and huddling, and, in those cases where the robots' behavior failed to match that of the real rats, it was because real rats can bend in the middle and the robots couldn't.

We see a similar effect if we return to the didabots. If you remember, the sensors for detecting objects were placed at an angle on each side of the didabot. This is what allowed them to avoid objects to the side, but to push those in front of them. Changing the didabots' bodies by moving one of the sensors so that it faced straight forward changed their entire behavior. Now, when they encountered an obstacle directly ahead, they would avoid it, not push it. Without the pushing behavior, no clustering takes place. Altering the position of the sensors alters the apparent "goal" of their behavior in the arena, even though the single "rule" with which they had been programmed remains completely unchanged. In all these cases—insect walking, pup huddling, and didabot clustering—the unavoidable consequences of possessing a certain kind of body lead to the emergence of adaptive and highly effective behavior, via environmental interaction.

Taking the Strain for the Brain

> There is more wisdom in your body than in your deepest philosophy.
>
> —*Friedrich Nietzsche*

The way in which bodies can be used to off-load the costs of cognition is a topic that is being pursued with great vigor at the moment in the world of robotics and artificial life (AL). It brings in some of the principles we discussed in our Watt governor example, where real-world physical constraints offer an advantage, rather than act as a hindrance, and it also reinforces the point that it is a mistake to assume that brains are all that matter in the production of complex behavior: bodies can help "spread the load" (in terms of both energetic and cognitive costs) that otherwise would have to be borne by the brain.

Take our legs, for example. When we walk, the swing of our legs isn't controlled by signals traveling from our brains to our muscles and back again, but occurs "passively" through the exploitation of gravity, friction, and momentum. There are neat devices called "passive dynamic walkers" that help to demonstrate this principle.[25] They can walk down a slope, using a very realistic human-looking gait, but have no sensors, motors, or any form of control system at all, except gravity.[26] They are, in the words of one set of researchers, the "gliders of walking robots."[27] As long as the legs are constructed in the right way to ensure pendulumlike motion, and the feet are sufficiently springy, the walker will move down the slope propelled by gravity alone. Of course, if there is no slope, then the walker is stuck and can't go anywhere. Recently, a team of researchers built a bipedal robot, based on a passive walker architecture, with small power sources at its ankles or hips that give it the ability to walk on level ground. These robots produce the same lifelike gait as a passive walker, demonstrating that the gait isn't dependent on the use of gravity as the power source, and also that the right kind of morphology may be key to cost-efficient movement: the walking efficiency of these bipedal robots is around the same as that of humans, whereas Honda's humanoid Asimo robot (which is controlled by a complex control architecture, with a large number of motors) uses about ten times the energy of a walking human.[28] Passive dynamic walkers—and their bipedal robot cousins—illustrate perfectly that using the right materials and exploiting reliable environmental features (like gravity) can reduce the brain-based costs of a behavior.

Our knees are also examples of this kind of "morphological computation,"[29] where the physical and mechanical properties of body parts do work that would otherwise require specific neural control (I have to confess, I don't really like this term, as you might imagine, because it implies that all cognition is computation, and that the body is doing some of the "computing" for the brain, when it's doing nothing of the sort). Biomechanically speaking, our muscle-tendon systems are rather springlike and elastic (and this is true of all animals; the use of springs in the Puppy robots is no accident). When our feet hit the ground, the movement of our knees is not controlled by our brains or spinal cord, but results simply from the mechanical properties of our legs. The only part of the process in which the central nervous system is involved is adjusting the degree of elasticity in our muscles (this needs to differ according to whether we are walking or running). For the rest, the knee can take care of itself, which is (at least partly) why we can move swiftly and easily over uneven ground.[30] We are not like those humanoid-robots that move with stiff jerky movements (because they are not made of elastic materials), and that also do so slowly because their joints and leg movements are controlled by motors that require activation from signals sent from a central controller. In other words, differences between the gaits of real animals and those of robots stem from differences in the materials from which the robots are made, not from variations in programming skill or processing speeds.

Another example of morphological computation is revealed through examination of an activity where more neural activation is required than one would imagine. Specifically, it requires more neural activation for a monkey to move just one of its fingers than for it to move all four of them, plus its thumb, at the same time (as when we form a fist or make a grasping action). This seems counterintuitive at first because more fingers means more muscle movement, which means more motor nerve activation and so more neurons recruited to control their action. This commonsense intuition is wrong, however, because the muscles and tendons of the hand have evolved in such a way that they naturally come together as a single unit when the hand closes, which requires much less in the way of neural control; our fingers can't help but come together in just the right way when we close our hand.[31] The reason our hands should work this way becomes completely intuitive—if not blindingly obvious—when we consider primate evolutionary history. Grasping hands evolved in the earliest primates to aid effective movement through the trees, and perhaps also to catch nocturnal insects. Natural selection did not act on the brain and nervous system to produce controlled hand movements; rather, it acted by exploiting the mechanical properties of the tissues making up the hand, creating synergies between muscles and tendons that saved on more costly (and slower) central nervous system control. When conditions changed and selection favored the use of precision grips and the like, greater control could be achieved only if this automatic mechanical response were overridden. In other words, because monkey fingers naturally open and close as a single entity, extra neural activation is needed to inhibit and override this so that only one finger moves, and not all of them. The resulting system is rather clunky as a result, but that's the nature of the evolutionary process; it has to tinker with what already exists and cannot generate completely new solutions from scratch.

Similar mechanical principles are now being used in the development of more effective prosthetic limbs. The "Yokoi hand" is a robot hand that can pick up nerve signals from the muscles in the remaining part of an amputee's arm and use these to control the hand's movements.[32] The hand has tendons made of elastic and fingertips made of a material that deforms when in contact with objects and surfaces (a bit like our real fingertips). This gives the hand two advantages. First, it means that the hand "self-adapts" to the objects that it grasps, reducing the need for motors to control every specific movement the hand makes. This not only makes the arm lighter but also increases the range of objects that can be grasped, because the hand need only approximate the right kind of grip, and the deformable materials then ensure that the hand adjusts appropriately. The second advantage is related to the first in that only very crude signals from the arm are needed to control the hand, because the hand itself does most of the work "on the fly." The "computing" of an effective and stable grip is performed by the hand itself in its moment-to-moment adjustments to objects in its grasp.

Understanding the possibilities that exist for morphological computation again helps us to "put the brain in its place"—which sounds pejorative but isn't. Too often, we give the brain priority, even to the extent of reducing the body itself to the brain. Specifically, we reduce the body to the somatosensory strip that runs across the top of our brain—the neurons to which the sensory nerves from our body project and which, if stimulated, give rise to a sensation in that part of our body. The number and patterning of these neurons can be represented as a "sensory homunculus" showing the relative size of the sensory areas devoted to different parts of the body (you may have seen these pictures or models—they have huge hands, mouths, and tongues but tiny legs and arms). The idea that the body is "represented" in the brain in this fashion gives the impression that the body's only job is to provide the raw input to the brain, and then to execute the brain's "master plan." It suggests that our sensory functions are embedded in our brains, when it would make more sense to "evert the homunculus"[33]—turn it inside out—and view the homunculus as the extension of our brain into our bodies.

The standard view of an inward-looking homunculus embedded in the brain is what makes the "brain in a vat" thought experiment seem so plausible and appealing: this is the idea that, as long as there were sufficient input fed into such an isolated brain, we would have just the same experience of the world as we do walking around in our body (which, if you hold to the idea that most of the world we see has been constructed by the brain anyway, would make an awful lot of sense). Some have gone so far as to suggest that this isn't actually a thought experiment at all: "each of us is precisely a brain in a vat; the vat is a skull and the 'messages' that come in are coming by way of impacts on the nervous system."[34] Understanding how the body itself can help produce complex, adaptive, and appropriate behavior undermines this kind of argument completely. It is impossible for the experience of a brain in a vat to be like our experience of the world, if a large part of that experience depends on the physical structure of our bodies and how they encounter the environment.

The Eyes Have It

Another lovely example of the way that bodies can relieve the burden on brains can be found in the compound eyes of insects. Compound eyes are made up of many tiny facets, known as ommatidia, each of which acts as a separate visual receptor (they are long tubes with light-sensitive cells at the bottom). Compound eyes produce a pattern of dark and light dots, a bit like the image you get if you enlarge a photo from the newspaper. The more ommatidia there are, the less "grainy" the image. If you look closely at the compound eyes of many insects, like houseflies and dragonflies, you'll see that the arrangement of ommatidia over the surface of the eye is not even; the ommatidia are more closely packed together at the front of the eye (i.e., in the direction of travel) than they are at the sides. The reason for this arrangement is that it allows for the eye to compensate for "motion parallax."[35]

Parallax is the way in which objects seem to shift in position when they are looked at along different lines of sight. If you look at an object with only one eye open and then shift to looking at it with your other eye, the object seems to shift its orientation slightly. This is because your eyes are positioned so that each eye's visual field views objects at a slightly different angle. This allows us to use parallax to perceive depth and judge distance. Motion parallax is the depth cue (or invariant) we gain as a result of our own movements—when we move at a constant speed, objects that are closer to us appear to move past us at a much faster rate than objects that are more distant—and we can use this invariant to judge the distance of objects; sitting on a bus looking out the window, we know that a slow-moving tree is farther away than one that seems to rush past.

In a compound eye with an even distribution of ommatidia, motion parallax is inevitable: a distant point of light will take a certain amount of time to move across one ommatidium and on to the next as the insect flies past it. A point of light from an object that is close to the insect as it moves past will take much less time to travel the same distance across the ommatidia. An eye in which the ommatidia are densely packed at the front and spaced out on the sides, however, can completely compensate for this effect: as the fly moves, light from a distant source will take the same amount of time to move across the densely packed ommatidia at the front of the eye as light from a close object will take to move across the widely spaced ones at the sides. Close objects will therefore appear to move at the same speed as distant ones, and there will no cues of depth available.

Why is this useful? One suggestion is that it allows insects to fly at a constant lateral distance from an object in a very cheap and cost-effective way. With eyes that compensate for motion parallax, all an insect has to do is fly in such a way that it maintains a constant "optic flow" across its ommatidia (that is, fly in such a way that the time taken for light from the object to travel across each ommatidium remains exactly the same). If it does this, it will fly parallel to an obstacle while maintaining a steady unchanging distance from it. If the optic flow changes under these conditions, it can only be because the insect has deviated from its lateral parallel path. If the ommatidia were evenly distributed, however, this tactic wouldn't work because motion parallax would also be a source of change in the optic flow, and it wouldn't be possible for the insect to fly in a way that kept this constant. Under these conditions, the insect would somehow need to compensate for changes in optic flow that were due to motion parallax, and filter them out. This would mean that ommatidia in different parts of the eye would need to have differently tuned neural circuits, which would increase both the complexity and the energetic costs of the process. Instead, simple changes in the arrangement of the ommatidia allow the eye itself to remove motion parallax at its source.[36]

The suggestion that varying the morphology of the eye allows insects constantly to maintain a flight path parallel to an object has received strong support from experiments with the "Eyebot": a wheeled robot mounted with an array of movable tubes with light-sensitive cells at the bottom (and so functionally equivalent to ommatidia, although real ommatidia don't move).[37] The Eyebot was given the task of moving at a fixed lateral distance from an object, and was allowed to "evolve" by moving its light-sensitive tubes into different configurations. Impressively, it took only five hours for the robot to converge on the design of the insect eye, with most of its tubes clustered closely together at the front, and with very few at the sides. Its behavior also changed accordingly as it evolved the ability to maintain its distance from the obstacle ever more effectively.

The Right Stuff

> Learn to dance before you learn to program.
> —*John Granzow*

Another excellent example of how being made of the right materials can make all the difference to behavior is provided by another of the robots that feature in Rolf Pfeifer and Josh Bongard's book.[38] Stumpy is a dancing robot, built by roboticist Fumiya Iida, that can produce all kinds of weird and wonderful "dancing" movements. What's intriguing about Stumpy, however, is that his inventors decided to build their walking, dancing robot in such a way that only its top half is powered by motors; its bottom half is entirely "passive," with no capacity to move by itself.

Stumpy consists of two T-shaped structures mounted one on top of the other. Its "upper body" is an upright T, and its "lower body" is an upside-down T to which four wide springy feet are attached. The point at which the two Ts meet is Stumpy's "waist," and this joint allows the robot to bend its upper body from left to right, but it cannot turn it forward or backward. The top horizontal part of the upper T is also jointed (a bit like a "shoulder") so that Stumpy can swing its "arms" from left to right. So Stumpy is a brainless robot that has a quite limited amount of powered movement and then only in its top half. When you power up Stumpy, however, it dances around the room using a variety of different stomping and jumping gaits, when you would expect that, at best, it would simply stand there swiveling its top half around. The reason Stumpy dances has nothing to do with its brain (it doesn't have one) or how it is wired up, but everything to do with the balance achieved between the different elements of its body and the materials from which they're made.[39]

So a version of Stumpy made with a much smaller top half and no shoulder joint cannot produce any nifty dancing because the top half isn't heavy enough to generate sufficient force to get the springy feet moving and generating their own momentum. Equally useless are versions of Stumpy that have a full upper body and shoulder joint but lack springy elastic feet. In this case, the forces and movements generated by the upper body can't be compensated for, and so the robot simply falls over. One could try various kinds of adjustments to make up for these shortcomings: for example, by adding motors to the feet to allow them greater control and to overcome the lack of movement produced by a small upper body, or by adding more control over the upper body to counteract the rigidity of the feet and prevent tipping. These kinds of adjustments are all more complex and more costly than exploiting the intrinsic control properties of particular materials (springiness, friction, stiffness), and gaining this greater control at no cost. Stumpy's inventors were able to demonstrate that, simply by combining the right kind of shape with the right kind of materials, one can achieve controlled but flexible movements.[40]

Understanding a little more about "morphological computation" really hammers home the relevance and importance of the body when we think about how other animals might see the world, and how (if) they think about it. It reinforces the point that we have to consider the whole animal, and not just attempt to isolate internal cognitive processes, when trying to understand their behavior, because, as we have already noted, we run the risk of putting into the head what evolution may have "farmed out" to the rest of the body. Just as with the parable of the ant, we will overestimate the cognitive demands of a behavior, and overcomplicate our models and explanations of it.

Taking a whole body perspective is also very "ecological": once we understand the role and importance of bodily attributes like materials and shape in producing effective, adaptive behavior, we can more easily appreciate that bodies are not simply the means by which a reasoning brain can implement solutions to the problems of life; rather, as Andy Clark[41] puts it, we can see that bodies are resources that can be exploited in various highly adaptive ways, either to reduce the costs of behavior, to produce more effective behavior, or both. We become concerned with the relationships that exist within bodies as different kinds of systems work together in concert, and we move away from a view of the brain as "director of operations" and the body as its (literally) mindless servant.

Beating "Cartesian Disease"

This idea can take some getting used to, not least because there are so many aspects of modern Western culture in particular that promote the "I'm in charge" brain-based view. When I was a child, the comic *The Beezer* used to have an ace cartoon strip called *The Numskulls*, in which several large-headed skinny-limbed men were shown living in the head of "Our Man" and controlling his behavior. Luggy worked his ears, while Blinky controlled his eyes, and—most mind-bogglingly of all—Brainy worked his brain (!). While the Numskulls raise all kinds of fascinating philosophical conundrums (do the Numskulls themselves have even smaller Numskulls inside their heads, and so on in an infinite regress?), the point here is that, even from a very young age, many of us are exposed to the idea that the brain is all-important, and it can be very difficult to shake (and how much more so now, following the "decade of the brain" and all those lovely fMRI and PET scan images that show brains glowing in vivid color). Once again, this doesn't mean flinging ourselves to the other extreme and insisting that brains aren't important at all, and that we can do without them. That is, obviously, ludicrous. But it is important to restore a balance, I think, and to see brains as parts of bodies that together make up a whole organism, whose main aim is to produce behavior appropriate to its circumstances. A brain is just one of many resources available to an animal to achieve this, along with the other physical properties of its body and its environment, and discovering how all these aspects fit together in the service of an animal's ultimate goals is an immensely exciting challenge.

Soft Assembly

Another eye-opening and liberating aspect of an embodied approach is that it gives us a new perspective on individual differences in behavior. Variability across individuals is often seen simply as "noise" or meaningless variation (measurement error) around a central tendency. An embodied, dynamical approach views variability as a blessing and not a curse. This is because the behavioral patterns shown by different individuals will be geared to their specific bodily dynamics operating in specific environments, and so we should expect to see variability in performance because of the necessary interplay between current local conditions and the specific idiosyncrasies of an animal's body. Behavior may, in other words, be a process of "soft assembly,"[42] where a whole variety of local control factors effectively exploit specific local (often temporary) conditions, along with the intrinsic dynamics of an animal's body, to come up with effective behavior "on the fly," just as in the case of the Yokoi robotic hand. Importantly, this variability won't be random; it will be entirely predictable and completely explicable in terms of individual bodily dynamics. It will be data, not noise.

Let's just expand on this a little with a clear, everyday example so that we firmly grasp this important idea of soft assembly.[43] Imagine a bunch of people in their offices who wish to have documents printed at various points during the day. Now imagine that these people's computers are networked to a bunch of printers, each of which can potentially print off any given person's document. The task at hand is how to distribute print jobs among the printers so that all the employees can get their documents printed in a timely and efficient way. One means of doing this would be to have someone (or something—a central computer) that monitors the print jobs as they come in, checks which printers have queues, and then allocates jobs accordingly: a complex task because these elements will vary stochastically and somewhat unpredictably; in addition, different documents may require different kinds of printing (color, photographs, large size) and printers will have different efficiencies. The solution to the task as presented above accords with our classical cognitivist view:

perceptual input (print jobs that need printing) comes into the system, planning takes place, and "motor" commands are sent to the printers, which then carry out the central controller's orders.

An alternative means of dealing with the same problem is to soft assemble it. We can do this by having each machine that needs a document printed send out a "request for bids" from the printers. Each printer responds with an estimate of the time it will take it to complete the job. If a printer has the right features (e.g., color printing) or a short queue, it will "outbid" other machines. The print job is then sent to the printer with the best bid, and these local interactions among the different machines will automatically ensure that print jobs are allocated among printers in the most effective way.

Efficient scheduling of jobs then emerges from the posting and bidding interactions that occur locally between active machines. There is no central controller, and no planning, which means there isn't a single computer or person responsible for the system as a whole, so the system is more robust (in the central-controller model, if the central organizing computer fails, or the person fails to show up for work, the whole system collapses). The links to PCT should be obvious here as well. It is also more robust because if any one printer should experience a meltdown, the system as a whole automatically compensates simply because the printer no longer sends out any bids, and so will never be sent a job that it cannot complete. This robustness will be accompanied by a certain amount of variability, given the peculiarities inherent in the particular context (e.g., the needs of different people for print jobs over time), so if we were to compare similar systems in two different companies, say, we'd see differences in the scheduling systems that reflected the particular idiosyncrasies of the computers, printers, and people who worked there. What goes for printers and print jobs can apply equally well to how an organism "schedules" the tasks it needs to complete as it behaves actively in the world. The peculiarities of its body in relation to its specific environment will lead to patterns of variable behavior across time and across individuals that reflect this opportunistic exploitation of available local resources.

We can also link this idea of soft assembly back to Michael Wheeler's point about distinguishing truly dynamical systems from the subset of computational systems that we can consider to be dynamical. Differences in intrinsic body dynamics will lead to differences in the richly temporal phenomena that act as the pacemakers and rhythm-setters that couple an animal to the world, and so truly dynamical embodied systems should display more adaptive variability in behavior than do strictly computational systems, where there is more likely to be a "right" way to do things in this strict sense. In other words, what matters is not whether a particular static kind of ability exists, but the relative stability of behavior across time and particular contexts among different individuals. Development is one area of research that benefits particularly well from this change in emphasis, and we can look at some examples of dynamic soft assembly in some very odd creatures indeed: human babies.

Chapter 10

Babies and Bodies

> Babies have big heads and big eyes, and tiny little bodies with tiny little arms and legs. So did the aliens at Roswell! I rest my case.
> —*William Shatner*

It is easy to underestimate human babies. The *Onion*, a spoof newspaper, once ran a story headlined, "Study Reveals: Babies Are Stupid." Among other things, the article reported that babies were unable to avoid getting their heads trapped in automatic car windows, nor could they master the skills of scuba diving or navigate their way back to land from the center of Lake Erie using a nautical map. It's so funny because, as they say, it's so true, and it is, of course, unfair. Babies don't have the sensory or motor skills to carry out these tasks, and we laugh because we know this to be the case. What we're failing to appreciate, however, is that when babies are flinging their limbs around seemingly at random, or are bizarrely enraptured by the sight of their own hands or deeply absorbed by the act of shoving their feet into their mouths, they are learning a ferocious number of skills that far outstrip those needed for scuba diving. Pointless and daft as most of it may seem, babies' behavior is laying the foundation for their entire conceptual knowledge of themselves and the world in which they live.

Both of the great developmental psychologists, Jean Piaget and Lev Vygotsky, recognized that action in the world was the initial source and cause of what, ultimately, ended up in our heads. This isn't the place for any kind of in-depth summary of their work; suffice it to say that (a) the details of how Piaget and Vygotsky envisaged this process differ (although less so than most developmental textbooks would have us believe), and (b) not everyone agrees with their views on development and cognition, and some would say that certain aspects have been discredited completely. For our purposes here, it seems fair to say that they both saw embodied action as essential to the development of a conceptual understanding of the world. After all, what other means do we have? How else can something more (an adult) be made from something less (a baby)? One option, of course, is that we are simply born with all the knowledge we need, but, as we saw in chapter 3, wholly innate mechanisms would be either too loose or too highly specific to allow for truly flexible behavior.

A more "constructivist" perspective, where each individual learns to "inhabit" its own particular body and builds knowledge up from what that body can achieve in the world is both more forgiving and more flexible in terms of coping with novelty and change. As with the face perception studies we discussed in chapter 2, a set of experience-expectant biases from which more complex knowledge can be built—knowledge that is tailored to the specific world and body we encounter—can explain our flexibility, while the fact that all humans share a particular kind of embodiment can explain why, in general, we all see the world in our distinctively humanlike way.

Learning to Live in Our Bodies

> A two-year old is kind of like having a blender, but you don't have a top for it.
> —*Jerry Seinfeld*

Shaun Gallagher—a philosopher who has spent a long time thinking about how embodiment shapes our minds—argues that one of our experience-expectant biases (although in this case it is something more than a bias) is our "body schema":[1] the system of sensorimotor capacities that give us our sense of our bodies in space, and how the parts of our body are located relative to each other. Close your eyes and raise your arm: you will know exactly where your arm is, even without looking at it. Similarly, if you stand up and then raise one leg, your body automatically adjusts; a variety of muscles in other parts of your body contract so that you remain upright, but you aren't aware that this adjustment is happening. This is because the body schema functions without any conscious awareness or the need for visual monitoring. Gallagher calls this our "proprioreceptive self," proprioreception being the name for this ability to locate ourselves and body parts in space. The body schema is distinct from our body image, which is composed of the conscious perceptions, attitudes, and beliefs we hold about our bodies. As I type these words, my fingers fly across the keyboard as I look at the screen and I somehow just know where my fingers are; I don't need to work at it or be conscious of what they are doing. I can, of course, choose to perceive my fingers in the act of typing, and bring them into perceptual focus, but then the accuracy of my typing immediately goes to pot. My body schema is involved in the accomplishment of my typing motions, and my body image is involved in perceiving that motion if I so choose. The important point, as far as Gallagher is concerned, is that our body schema fundamentally shapes the structure of both our unconscious and conscious thought processes. It does so because it is the center of our perceptual field. Our bodily awareness is the reference point for all our worldly experiences, built right into our structures of perception and action, and this awareness of our body is just "given." We have a sense of our body at all times, but this doesn't depend on our consciously viewing our body as an object (if it did, we would soon find ourselves in an infinite regress, like the little man in the head who looks at the image on the retina).

Our body schema is innate, at least in rudimentary form. A baby has some sense of its own body and uses this to control and coordinate its movements. The spontaneous movements of the baby in the womb appear to shape the body schema—so although innate, it is not hardwired. The proprioreceptors in our muscles appear at 9 weeks of age, and a fetus will show spontaneous, repetitive movements soon after. At the age of 12 weeks, a fetus will also start making repetitive hand-to-mouth movements. As Gallagher notes, these movements may help to generate and facilitate the development of our body schema.[2] What this further suggests is that, although our body schema shapes us according to the same general plan, so that we all turn out roughly the same, it doesn't mean that we all get there in the same way.

This is because all bodies are different. We learn how to exploit our specific bodily resources, and that requires "customized" learning strategies and developmental trajectories. Studies of babies' stepping behavior illustrate this perfectly. When they are first born, babies spontaneously make stepping movements when they are held upright. This behavior vanishes when babies reach about 2 months of age, only to reappear again during the second half of their first year when they bear their own weight on their feet. Why such a strange pattern? Traditionally, it was argued that very early stepping was an involuntary, somewhat "random," motion that disappeared once the nervous system had matured sufficiently to inhibit these actions. When stepping reappeared, it was in a more controlled intentional form, designed for walking.[3] Another (more cognitivist) suggestion, based on findings that babies who were made to practice stepping frequently continued to do so beyond 2 months,[4] was that these early primitive motions could be "captured" by higher-order learning processes, and so become voluntary as a result. The reason that babies stopped stepping was that, under natural conditions, they didn't "practice" in the same way, and so the skill was lost owing to disuse.[5]

In both these cases, the change in the intentional nature of stepping (the switch from involuntary to voluntary actions) that seems to accompany its loss and gain was assumed to be the single underlying cause of the process. But early infant stepping, while involuntary, isn't a random flailing of the limbs; rather, it is the well-coordinated, alternating flexing and extension of the hips, knees, and ankles, all at the same time, just like "proper" controlled walking. What is more, babies also show the same movement pattern when lying on their backs and kicking: if you pick up a kicking baby, it begins stepping. "Kicking" and "stepping" are not distinct processes, just different names for the same actions performed in different positions. This distinction is more than just an issue of semantics. Unlike stepping, kicking isn't lost after 2 months but continues throughout the first year of life. Stepping also declines faster in babies that have gained the most weight between 2 and 6 weeks old. If stepping is simply a matter of brain maturation, why isn't kicking affected in the same way, given that this is simply stepping horizontally? And why should a baby's weight be relevant?

By adopting a dynamical view with an emphasis on self-organization and soft assembly, Esther Thelen and Linda Smith were able to come up with an alternative explanation as to why stepping vanishes at 2 months.[6] Put very simply, their suggestion was that the mass of a baby's legs was more important than the maturity of its brain. By 2 months of age, an infant's legs have simply become too chunky to overcome the force of gravity. We can show this experimentally: by placing weights on stepping babies' legs to tip them over the heaviness threshold, one can prevent stepping babies from producing these movements. Holding nonstepping babies in water (which supports their legs), by contrast, leads them to spontaneously begin stepping again. Babies who were given "practice" at stepping may have continued to do after the age of 2 months not because of any reorganization of their brains, but because of the building up of their muscles.[7] Specifically, the babies in the experimental group were held up to step, while those in the control group had their legs moved around while they lay on their backs. Holding babies upright places "overload" demands on their muscles by making them work against gravity (similar to the way adults work against gravity when doing sit-ups), and this increased resistance may have strengthened their legs sufficiently to enable them to overcome the greater mass of their legs as they grew. This may also explain why children in a number of non-Western cultures retain stepping behavior beyond 2 months and also walk earlier;[8] like the babies in the experiment, they are "trained" by their mothers and caretakers to adopt various kinds of upright postures and engage in various movements that help strengthen their legs.

Other experiments showed that, when 7-month-old nonstepping babies are placed on a treadmill, they produce perfectly alternating, coordinated stepping movements, and can do so even when placed on a setup with two parallel treadmills that move at different speeds. Babies can do this despite the fact that stepping behavior is completely involuntary, just as it is among young babies who have yet to lose the stepping behavior: the stepping babies don't look at their legs or pay attention to their movements when on the treadmill. Instead, it seems as though their legs are just able to adjust naturally—something that immediately should lead you to suspect some kind of morphological "computation." What seems to happen is that the treadmill is instrumental in helping to soft assemble stepping: the motion of the treadmill pulls the baby's leg back until it reaches a point where it can't go back any farther and it suddenly springs forward again—much like what happens when you stretch a spring and then let it go.[9]

These synergies between muscle and tendon in the legs are the reason why babies don't need to "think" about what they're doing on the treadmill, in any classic sense; instead, their movements self-organize, thanks to the structure of their legs, the structure of the treadmill, and the interactions between them. This is why nonstepping babies can begin stepping in such an insouciant manner: the behavior has as much to do with the environment as with the baby. This means that we can, as Andy Clark notes, treat the environment as an "equal partner" in producing behavior that is softly assembled in this fashion.[10] Once again, we can see why softly assembled systems are often more robust and flexible than systems with a "classical" centralized control architecture. Whereas a centralized control system may break down completely as a result of a novel change in the environment (because it hasn't been specifically programmed for that particular environmental contingency), a softly assembled system can easily adjust because the environment is already part of the system and so is partly responsible for determining behavior. It becomes more difficult to catch a softly assembled system completely "unawares."

A dynamical soft assembly approach to development also helps to makes sense of the wide variability seen across babies in when they learn to walk and how they do it. If it were just a matter of their brain systems maturing, we should expect to see more uniformity in the when and how of walking. Once we understand that babies have to learn how to exploit the contingencies and resources of their own particular bodies, however, it becomes obvious that variability should be the norm: a chubby baby whose legs are heavier will show a developmental trajectory that differs from that of a smaller or lighter baby. That is, although chubby and skinny babies will both show the same overall trajectory—stepping→no stepping→stepping—if the trajectory is measured in terms of the babies' chronological age, and they are then compared, the specific patterns of timing will differ between the two (the chubby baby will stop its initial stepping sooner, for example).

Studies of how babies learn to reach for objects successfully show similar patterns. A highly active baby that flings his arms widely in an energetic fashion is more likely to bat an object farther out of reach than to grab it. To reach for objects effectively, these energetic babies must learn to slow down their movements and bring their hands into a position where they can see them, so that they can visually guide them. In this way, objects are contacted at lower speeds, and the babies' grasping movements become more coordinated. Passive, quiet babies, in contrast, must learn the opposite: they must become more energetic, and move around more, so that grasping of any kind can become a possibility.[11]

A baby has to explore and learn about the possibilities of its own body and the movements that it makes, with the result that it develops a highly personal understanding of the world, constituted by the ways in which the "richly temporal" rhythms and pace of its own body constrain and enable certain actions and processes. Development isn't an "inevitable march toward maturity" but something more fluid: a process that unfolds over time, showing patterns of dynamic stability that evolve and dissolve, as the result of the varying interplay of multiple internal and external influences.[12]

Crawling, Walking, Reaching, Thinking?

> Thought grows from action and that activity is the engine of change.
> —*Esther Thelen and Linda Smith*

> Ever tried. Ever failed. No matter. Try again. Fail again. Fail better.
> —*Samuel Beckett*

We can see this by looking at how babies tackle the transitions between crawling and walking. Crawling is what babies do to move themselves around when they have enough strength to get up on their hands and knees and coordinate the movements of their arms and legs, but not enough strength to balance upright and move at the same time. Crawling is also a "self-organizing" behavior: it's not built into us that we should crawl (that is, the baby doesn't have the "goal" of crawling as such, although we can attribute that from our outside frame of reference): it's just the most stable state that the dynamical system—or, as we usually call it, the baby—will adopt to get itself across a room, given the affordances offered by its body and the environment. As babies get stronger, they make the move to walking, and that has the effect of "destabilizing" the crawling behavior, because there are new behavioral patterns influencing ("perturbing") the system and influencing its dynamics. Crawling doesn't teach babies how to get about in the world in any kind of general sense. Instead, the process is more like the distorted lens experiment: crawling babies have learned about the specific affordances of their body and environment, and how to exploit them, in relation to crawling. What they haven't learned is a set of generalized abstract rules about what makes for efficient locomotion.

We can see this by looking at what happens when crawling versus walking children are placed on a steep slope (20°).[13] Toddlers that have just begun to walk[14] are often wary and fearful on a slope (as well they might be, given the high chance of falling over). Many children will simply refuse to tackle the slope or will switch to a different means of locomotion, like crawling or sliding down headfirst in a "Superman" posture (basically, lowering their center of gravity and making themselves more stable). If one takes crawlers and places them on steep slopes, however, initially they will plunge down the slope at full throttle and will usually fall. Unlike walkers, who have experienced falling over a lot as they began to walk, and can detect the affordances of stable versus unstable surfaces, crawlers know no fear. After a few experiences with a slope, however, they, too, can learn to detect its affordances and can learn to avoid steeper, riskier slopes on which they will topple over, and discover effective strategies for moving down less risky ones.

If you wait for these experienced crawlers to begin walking and then try them out on a steep slope again, a remarkable pattern is seen: "slope-experienced" new walkers will plunge down risky slopes at the same rate as they did when they first encountered them as crawlers.[15] All the weeks of learning to deal with steep slopes and to avoid impossible ones do not have any influence on how they respond to the slopes as walkers, and they don't learn any faster the second time around. Everything these children learned about slopes seems to have been lost, and, in a very real sense, it has, because their knowledge is tied fundamentally to their posture. Indeed, their knowledge is so tightly linked to posture that babies show no transfer of their skills on a slope from one trial to the next: if new walkers are placed into their old, familiar crawling posture, they either refuse a risky, steep slope or slide down it; if they are then tested again immediately on the same slope but in a walking position, they attempt to walk down it and have to be rescued by the experimenters! So babies don't "know that" one falls down steep slopes because of angles and centers of gravity and balance. Rather they "know how" to deal with slopes; they have learned to perceive the affordances that a slope offers to crawlers, and, via self-organizing processes, they have acquired the perception-action loops needed to successfully deal with them. Making the transition from being immobile to crawling to walking involves a whole new suite of perceptual inputs and changes in bodily dynamics; a child therefore learns to perceive these new affordances and performs new actions that exploit its bodily resources in completely new ways.[16]

This may also help to explain the pattern of errors babies make when given some classic cognitive tasks. There is, for example, a test called the "A not B test."[17] A baby is shown an enticing toy that is then placed in one of two containers directly in front of the baby. After a short delay, the baby is allowed to reach for and retrieve the toy. After several rounds of the toy's being placed in container A, the experimenter switches location and places the toy in container B instead. Jean Piaget—who devised the test—found that 8–10-month-old babies would continue to reach for container A, despite watching the toy being placed in container B, whereas babies of 12 months or older didn't make this mistake. Piaget thought this was because younger babies didn't have an "object concept," so they didn't fully understand that the toy existed independently of their own perceptions and actions.

One can introduce slight modifications to the test, however, that allow younger babies to pass with ease. Removing the delay before the baby is allowed to reach eliminates the error; so does putting weights on the baby's arms or changing its position (from sitting to standing) as it watches the toy being placed in B, and during the delay before reaching. These findings don't sit well with the highly cognitive explanation offered by Piaget, but they do lend support to a more dynamical interpretation. Put simply, the idea is that the reaching system of the baby has to be perturbed so that it doesn't fall into the attractor basin created by repeated experiences of reaching for the toy in container A. When babies reach for A in the first few trials, the memory of each reach becomes another input to the reaching system in the next trial—activation for location A is increased with each subsequent trial because of the previous activity that took place there—and leads to the formation of an attractor. Although babies receive a strong cue to location B from the experimenter when the toy is hidden in this new location, this decays during the delay between hiding and reaching. The stronger memory (or attractor state) of its actions toward location A comes to dominate, and the habitual reaching response is made.[18]

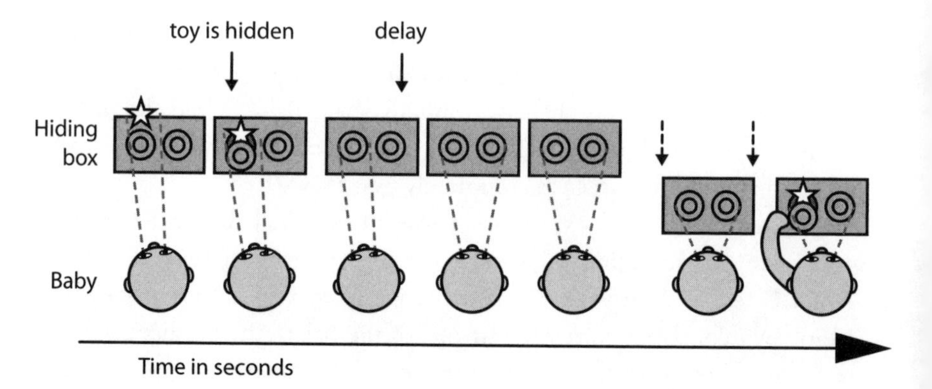

Figure 10.1. The design of Piaget's classic A not B test. A baby is shown a toy, which is then placed in one of the containers, A. There is a short delay, and then the baby is allowed to reach for the containers and retrieve the toy.

This is why removing the delay works: it allows babies to reach for B while the cue is still "active." Disrupting the physical dynamics of the system by weighting the baby's arms or changing its posture plays a similar role. If the relevant memories for previous reaching are found in the body itself—in the sensorimotor coordination of the babies' arm movements—then changing the baby's posture will destroy those memories. When the baby's arms are heavier or she is now standing rather than sitting, the baby's bodily dynamics are changed—it's more difficult for her to move her hands, or they are now in a different location relative to the target—and so different movements will be needed for effective reaching. This leads to changes in the dynamics of the system as a whole, allowing them to "escape" from the reach-for-A attractor basin.

The manipulations introduced by the experimenter help younger babies pass because they shift the dynamics of the system in ways that children can't yet achieve on their own. As babies grow older, they are able to shift the dynamics of their reaching systems without any outside help. This is where crawling and walking come in. One suggestion is that, as babies grow stronger and can begin to move around on their own, they encounter the same objects in a variety of different ways—on their knees, standing, from the front and rear, above and below—and their bodily dynamics must adjust to these differences with each encounter.

This broadening of their experience—and the baby's own role in shaping it—may make it easier to create and sustain new attractors and shift between them.

Doing without Concepts?

One of the major points of Linda Smith and her colleagues' study of the A not B test was that their explanation has no place for an "object concept" as conceived of by Piaget.[19] The task for babies is to reach to the right location in visual space: they must generate a motor "plan," maintain it over the delay, and then put it into action. Smith and Thelen's radical idea (which remains controversial) was that this motor plan is part and parcel of the "belief" that objects persist in space and time: it is, if you like, an action-oriented representation. In their view, one cannot say that a 12-month-old differs from a 10-month-old because he or she has gained a static "object concept" and that is the single cause of the differences in their behavior. Rather, their argument is that the babies' reaching is softly assembled in-the-moment with many contributing forces that make the error appear and disappear. The "belief" resides in the system as a whole—the baby coupled to the world—and not in the baby's head as some kind of symbolic, static representation. As Linda Smith puts it, "Cognition just is an event in time, the emergent product of many heterogenous systems bound to each other and to the world in real time."[20] As we've already noted, however, such a position has a problem explaining "representation-hungry" processes. (This dynamical systems view has also been criticized from a developmental perspective. Other studies have shown correct performance by babies on tasks that are very similar to the A not B test, except that they do not require babies to reach actively but simply look in the right place. Results like these have been argued to show that babies do have an "object concept," but that active reaching somehow disrupts children's ability to display this competence.[21] As Linda Smith notes, however, it is almost impossible to disprove the idea that babies have an object concept that doesn't have to actually show itself in behavior; basically, the argument is that we know it's there even though we can't see it, which isn't particularly sound as arguments go.) This may matter less from our perspective because we are applying these ideas to nonlinguistic animals (of which babies are a good example, at least initially) where such criticisms do not apply. For us, the idea that "beliefs" reside in and are distributed across sensorimotor processes in the activation of a motor plan is immensely appealing (and, of course, even if there are some "constants in the head," this doesn't deny that they may be distributed across sensorimotor processes as well). It doesn't reduce animals to "empty boxes" or simple stimulus-response machines, nor does it require us to place concepts into the head of the animal that may have no "real" existence. In this view, behavior is an actual constituent of a mind, and not simply an outward clue to what the "hidden mind" is doing, allowing us to avoid contracting "Cartesian disease."[22]

Multimodality Is Marvelous

> If you hold a cat by the tail you learn things you cannot learn any other way.
> —*Mark Twain*

The way that babies learn how to cope with slopes, or solve reaching problems, is a microcosm of how all our knowledge about the world—and our place in it—develops. As babies perceive and act in particular contexts, they are forming a multitude of "time-locked" multimodal correlations that link together their various sensory modalities and how the information they pick up from the world is affected by the ways in which they move and act.[23] "Time-locked" simply means that when you hold an apple, say, you're acquiring information on its color, seeing that its skin shines in the light, and how this changes as you move it around. At the same time, you feel its texture: how smooth it is under your fingers. A crawling baby on a slope is picking up how the texture of the surface feels, how its arms and legs are positioned, and what the world looks like from this particular angle. All these different kinds of sensory experience become "locked together" in time, and, with greater experience, these correlations allow us to generate complex patterns of sensorimotor coordination (and, eventually, depending on your take on these things, abstract concepts). For example, over time, correlations between the feel and sight of an object can become decoupled: we don't need to touch an object to know what it will feel like; we can categorize it just by looking at it. The presence in the brain of "canonical neurons"—motor neurons that fire both when we grasp an object and when we simply look at that same object—provides neurobiological support for this "decoupling" idea. Our perception of an object is not simply perceptual but also, to some degree, "conceptual" we don't just see what an object is; we also immediately "see" what we can do with it.

The other consequence of "time-locked" learning is that different sensory systems can "educate" each other about the world without requiring some kind of central mechanism (a "teacher") to coordinate everything (soft assembly again!). As you play with an object, turning it over in your hands, your visual and touch (haptic) systems automatically educate each other about what the object feels and looks like. There are four different kinds of "mapping" that take place during this process. One between the object and activity in the visual system, one between the object and the haptic system, and two "reentrant" maps: the activity in the visual system is mapped to the haptic system, and activity in the haptic system is mapped to the visual system. In this way, the visual quality of, say, smoothness, is correlated with the feeling of smoothness.[24] As you manipulate the object and notice changes in how it feels—maybe there's a raised edge or a rough patch—you are also picking up on changes in how the object looks. The two different experiences inform each other via the reentrant maps. When babies do seemingly tedious and "mindless" things—gazing in wonder at their own hands waving about in front of them for minutes on end—they are learning how the changing views of their hands correlate with how their hands feel as they move them about in space. To put this in linguistic terms, it's as if they were saying to themselves, "Hmm, when I get this flappy feeling in this part of my body, I also see a fast-moving, blurry object in front of my face." Over time, they come to realize that the flappy feeling and the blurry object are their own hand in space.

Experiments on very young infants have demonstrated the potential power of reentrant mapping.[25] Babies were given either a standard smooth dummy (pacifier) to suck on or one covered with lots of little knobs. When presented with the two different kinds of dummy visually, the babies looked longer at the kind of dummy they had just been sucking. The longer looking time is taken to indicate higher interest, so the inference is that babies are more interested in the previously sucked dummy because they recognize it visually, even though they had only touched it.[26]

Findings like these have been used to argue strongly against an "indirect" view of perception, and the idea that we have to integrate the different sensations coming in from the separate sensory modalities to generate a coherent representation of the world around. Instead, they suggest that, from the very first, we are inherently multimodal creatures; our different senses are already and always communicating with each other, via loose coupling through the environment.

The ability of infants to understand transparent objects is a good example of this kind of multimodal learning. Babies with no experience of transparent objects perform very poorly when they have to retrieve an object from a see-through plastic container. Instead of reaching around and into the container, they will try to reach directly through its solid walls. This isn't because they don't understand about containers or how to get objects from them: babies have no problem reaching around and into an opaque container in order to retrieve something. What they lack is the recognition that transparent objects, while see-through, are also solid; their lack of experience means they haven't been able to use their sense of touch, which tells them there is an object present, to educate their visual sense about the specific visual properties of transparency (transparent objects aren't "invisible" after all; they reflect the light in ways that allow us to recognize them as solid). If babies are given experience with transparent objects—handling them and playing with them, with no overt form of instruction or guidance by their parents—they can retrieve objects from transparent boxes with no problem at all. They have learned to perceive the visual cues that indicate the presence of an object that will be solid to the touch.

Constraints and Opportunities

One needs very little to get the ball rolling on this process. In principle, even random neural activation will allow babies to explore their bodies and the world in productive ways. Again, this comes down to appreciating how the body, the brain, and the world interact. Because our anatomy tightly constrains what movements are possible, random neural activation doesn't lead to random movement.[27] Thanks to the skeletal structure of our arms and shoulders and the ways our muscles and tendons work together, our arms are more likely to swing forward than backward, with the palm of the hand facing in toward the body. Random sparking of an arm movement is more likely to cause a baby to swing its arms in front of it than behind it. If, as a result, the baby contacts an object and grasps it, again, the greater likelihood of forward movement brings the object into the baby's visual field, presenting the opportunity for correlating visual and tactile experiences. The swing of the arm is also likely to bring an object near the baby's mouth, allowing its taste receptors to get involved, bringing in another sensory correlation. As it does so, all the different sensory experiences are tied together by the fact that they occur simultaneously in time, and the baby learns not only about how to coordinate and exploit its own body, but also about the object's size, weight, color, and texture. In this way, a baby solves the "learning paradox": how does someone who doesn't know what should be learned and what should be ignored actually manage to learn something useful? Their actions, even if random initially, present them with tasks to learn (how to grab something in their field of view) and the means for doing so. As they learn, these sensorimotor correlations form the platform upon which more sophisticated knowledge about the world can be built, all of which is precisely tuned and calibrated to the baby's specific situation and context.

Sensorimotor coordination of this kind has also been used to allow robots to detect the presence of objects in the world. One of the challenges of robot vision is to find a means of detecting objects against a background (and of course this is a challenge for real animals too). As we discussed in chapter 6, Gibson argued that animals use their own movements as a means to detect invariants and so pick up information about objects in the world. The Babybot takes this idea a step further by using sensorimotor coordination as a means to detect the presence of objects. The Babybot has a motion-detection system that allows it to "see" the movements of its hand and arm in space. This is all it can see—it isn't programmed to see the environment beyond its hand and arm. It can come to learn about this environment and what is in it, however, by contacting objects with its hand and arm. When this happens, it causes an immediate spread of motion activity in the Babybot's field of vision, which is larger than that of the arm alone, and this spread of motion activity is very easily detected. In this way, the object becomes "visible" against the background. The robot then "knows" what part of its visual field is constituted by the object. It should immediately be apparent that this is an example of the kind of sensorimotor coordination that Dewey spoke about, and that it also has shades of Gibson's ecological approach, as the robot can detect an object in its environment only by actively exploring its environment and acting on it, by poking things. One can also see how this can make the task of learning easier, because the robot's motion provides a kind of "filter"; it isn't overwhelmed with sensory input because it "sees" only those objects that it moves by its own actions.[28]

Our embodied, physical knowledge of objects persists throughout life. Even once we have acquired language and the ability to form abstract, symbolic concepts of objects, our knowledge of them is still fundamentally grounded in the actions we can perform on them and with them. In one study, subjects were asked to detect whether or not a visual stimulus of a pitcher had been presented to them.[29] Subjects responded by hitting a response key. This was all they were asked to do: a completely straightforward, visual discrimination task. Yet subjects' responses varied depending on how the pitcher was oriented with respect to their response hand. If the pitcher's handle was shown on the same side as their response hand, subjects were much faster to act and hit the key; they were considerably slower if the handle was on the opposite side from their response hand. This suggests that the action-centered representation of the pitcher, which included the possibility of grasping the handle, introduced an interference effect and slowed the subjects' response. This is the basis for the argument presented at the beginning of chapter 7 for why we metaphorically structure our worlds in particular ways: our metaphors are grounded in our physical actions. Some researchers go so far as to suggest that even the most abstract conceptual knowledge we possess, that of mathematics, is ultimately grounded in the ways our bodies deal with the world (which raises a lot of uncomfortable questions about the ultimate objectivity of the physical sciences: if our knowledge depends on having bodies like ours, can our findings really be "revealed truths" about the universe?).[30]

Degeneracy and Redundancy

The embodied multimodal nature of learning about the world relates directly back to our discussion of the umwelt and behavioral flexibility. We mentioned briefly how redundancy increases behavioral flexibility: that is, the more sensory modalities there are, and the more cross talk and education between them, the more "redundancy" there is in the system, which then allows for greater "degeneracy." That is, any single function can be carried out by different configurations of modalities acting in concert, and/or each configuration of modalities can participate in a variety of functions. We learn about the world using all our senses all at once (vision, smell, hearing, touch, taste, and balance) and, in this way, the information we obtain from the world via one modality partially predicts the information that can be extracted by the others. This is also a point that Gibson made with respect to perceptual systems: as the different modalities are overlapping, the same information can be picked up by a combination of perceptual systems working together, as well as by one working alone, so that, by constantly changing the input to all these systems at the receptor level, organisms can learn to discriminate and isolate the invariants at the level of the perceptual system.

To some degree, it is not strictly necessary to have all of these systems operating together in order for an animal to function effectively, but having this degree of redundancy provides greater robustness and flexibility. Although the information that each modality provides is overlapping, each one is based on different physical properties (vision is based on electromagnetic waves; touch is based on mechanical pressure), which means that, if any one of them should fail, it is not catastrophic: we don't grind to a halt completely if we lose one of our senses; instead we experience "graceful degradation"—we may lose a specific skill, but the system as a whole remains functional. Nor do the modalities themselves have to fail for degeneracy to be useful. After all, vision works only if there is light by which to see, whereas we can use our sense of touch regardless of lighting levels, and thanks to the way in which our different modalities have educated each other, we generally know (or at least have a very good idea) what something will look like just by feeling it.

Using several modalities is like wearing both a belt and braces to keep your trousers up: braces are redundant if you're also wearing a belt, but if your belt buckle should suddenly fly off, your trousers and your dignity are maintained at the appropriate levels. Redundancy is not without its costs, however. Possessing more than one way to extract the same information from the world is costly in terms of building the body parts needed and the energy they consume. This is why, unlike the redundancy in many human-built mechanical devices, biological redundancy is not always completely overlapping. This is an important point to note because while we are familiar with the fact that we humans have exceptionally large brains, it is also true that, we have very sophisticated and wide-ranging sensory and motor systems as well. Without this level of sophistication in our sensorimotor system, our big brain would be pretty useless; it is the balance between our bodies and brains that is instrumental to our success as a species. Think of it in artificial terms: a fast computer processor is no use to a robot if its body allows only for slow deliberate movements. The speed of the processor can never be realized in any useful way (say, by being used to detect and catch fast-moving prey), and so it just goes to waste. Without sophisticated peripherals, there is no point having a fancy processor.[31]

Babies and Other Animals . . .

We have spent a lot of time talking about our own species and artificial "species" like robots, rather than other animals, but, of course, if embodied processes are crucial to how we come to understand the world, then similar processes will be relevant for understanding other species as well, and for adopting a similar stance when we are investigating their cognition. For example, a more dynamical, embodied approach has a ready—and more interesting—explanation for individual variability, as we've seen. This is relevant from our point of view here because studies of, for example, monkeys, great apes, and the crow family also show plenty of individual variability in the level of performance on different types of cognitive task. Now, one can easily imagine that individuals will differ in how "smart" they are in terms of some general (rather nebulous) concept of intelligence, but it is also worth considering how factors relating to body size and strength could play a role, whether postural changes influence performance success, and whether the design of an experiment and its physical setup allow some individuals to exploit morphological computation effects more effectively than others can, or to capitalize on the physical affordances of the environment more effectively.

These can still be regarded as differences in individual "intelligence," but we would have to reconceptualize this in terms of the animal as a "complete agent" embedded in its world. As a result, we would need to pay more attention to how different animals actually performed on a task, rather than whether they simply passed or failed. This could allow us to get a lot more out of our experiments, because failures would become as least as informative as passes, and individual differences would be crucial to identifying the key to the processes used to engage with and solve problems. It would also leave us more open to those "odd" results, the serendipitous discoveries, that often leads to major scientific insights. Further, a more dynamical, embodied approach allows us to generate simple testable explanations for behavior without immediately succumbing to the "Cartesian trap" of attributing specific concepts that shape an animal's goals, which the animal then attempts to achieve by making and executing a plan.

As Linda Smith points out, concepts are, in a sense, "mythical": no one has ever seen a concept; they are theoretical constructs that we use to explain a phenomenon. If concepts don't add value to our explanations, and we don't need them to account for what we see, then we should get rid of them. The role played by sensorimotor coordination, morphological computation, dynamic coupling, and soft assembly can account for many of the flexible, adaptive, richly temporal behaviors that characterize living systems without any need for the concept of a concept. The very fact that I can write that last sentence—and have it make sense—suggests that we do need the concept "concept" to help us make sense of the world, but we also need to appreciate how firmly our concepts rest on a linguistic foundation. Words—and the beliefs, desires, and thoughts they capture—are context free and timeless in a way that sensorimotor representations are not; on a day-to-day basis, the meaning of a word doesn't change with either our physical and emotional state, the time of day, or the environmental context (although not everyone would agree with this position).[32] In contrast, and as we've seen, a baby's "concept" of an object (its belief that it occupies a certain position in space) is embedded firmly in its sensory and motor processes: the belief doesn't "mediate" between perception and action; it simply is perception and action. As a result, the concept of an object does change with context, or the physical and emotional state of the infant, so it isn't really a concept as we usually understand it, and it doesn't do any particularly useful work for us as we attempt to understand what the babies are doing. The same is likely to be true when we consider other animals that similarly lack symbolic language.

Language as a Cognitive Resource

> She had lost the art of conversation but not, unfortunately, the power of speech.
> —*George Bernard Shaw*

Language changes everything and makes concepts possible because, as Andy Clark has suggested, a thought in words is a bit like an object (metaphorically speaking), and an object is the kind of thing we can have other thoughts about.[33] According to Clark, language is, then, a kind of "double adaptation." He compares language to a pair of scissors, which also shows this kind of double adaptation. Scissors are well suited to the manipulative abilities of hands—they offer the right kind of affordances and are well designed for our particular anatomy (that is, a pair of scissor is basically screaming out to you, "Stick your fingers in these holes," and when you do that, the way your hand is positioned in the scissors naturally affords the movement of the blades up and down in a way that allows you to perceive their affordances for cutting things). Further, scissors have the added benefit of conferring on our hands abilities that they do not naturally possess: namely, the ability to make clean straight cuts in paper and other kinds of materials.

The same can be applied to language. It confers on us powers of communication and is well suited to the structure of our brains. But, at the same time, it confers on our brains powers that they don't "naturally" have: it can reshape difficult tasks into formats better suited to the capacity of the human brain. Clark argues that we are basically pattern-recognizers (which should be clear from our consideration of an embodied, embedded, ecological approach). Language allows us to "overcome" this default mode and tackle problems differently, problems that we couldn't manage without language to structure them. So Clark's point is that language allows us to "trade spaces." That is, by using the external symbol structures that make up language, we trade various forms of culturally achieved representation that take the strain off our brains. Language is not purely for communicating, but is also a way of effecting changes in our environment that enable us to achieve more than we could otherwise. So, when we write an essay, we are not having thoughts and then writing them down. Instead, the act of writing is itself an act of thinking, because it's a way of using language in a fashion that allows us to precisely order our thoughts and convey what we mean. The thought is produced by and through the act of writing. Without writing, we couldn't have these kinds of thoughts.[34]

It should be clear from this that Clark literally thinks of language as a tool. That is, it hasn't changed anything about the basic structure of our brains, or how they work, but rather it complements their functioning. It allows a parallel-processing pattern-recognizer to process things in serial fashion according to a set of precise rules. In other words, it allows the brain to operate "as if" it were a modern digital computer even though, as we've seen, brains don't seem to work in that way at all.[35] Language gives human brains a new way of dealing with the world. We haven't been "reprogrammed" by language; rather, we use it as we use a pair of scissors. It provides us with an extra loop of control over our behavior.

Clark also uses a very nice metaphor that he calls the "mangrove effect" to get this point across. Mangroves are strange tropical trees that can grow from a floating seed, which establishes itself in water and roots in shallow mud flats. The seedlings send vertical roots up through the water, and as they grow, you end up with trees that look as if they are on stilts. This complex system of aerial roots traps floating soil, weeds, and other debris. As time passes, this accumulation of mud forms a small island around the roots of each tree. This grows larger and larger, until the islands around a group of trees eventually merge. In other words, the land is built by the trees.

Clark suggests that words can sometimes be like this. While it is natural to think that most words are always "rooted" in a preexisting soil of previous thoughts, sometimes it works the other way around. Think of poems. Sometimes it is the choice of a particular word—its sound and structure—that influences the thoughts that the poem comes to express. Clark suggests that words can serve as fixed points, like the roots of the mangrove, that can position other kinds of intellectual matter, creating islands of thoughts about thoughts.[36]

By capturing the world linguistically in this way, we can, it seems, achieve the kind of "present-at-hand" stance that we spoke of earlier: a stance that allows us to detach and step back from the world, and so engage in the scientific pursuits that we need to get some kind of grip on how other animals might see and cope with their worlds. It may well be that part of what makes us different from other animals was driven by the evolution of language as an external tool that gave us the ability to create islands of "second-order thought" (thoughts about thoughts), with which we could then create and stabilize abstract ideas and concepts, including the concept that other animals' minds are very similar to our own. The irony here is that our very ability to entertain this concept actually exposes the difference between us and other animals. Animals lack the islands of words that can help stabilize and support such a strange and marvelous concept as a "mind."

Chapter 11

Wider than the Sky

I not only use all the brains that I have, but all that I can borrow.
—*Woodrow Wilson*

Last summer, my friends Shellie and Stefan and their son, Oliver, went on holiday to Italy. While in Rome, they decided to take a Segway tour.[1] Stefan became an immediate devotee, and while enthusing about its potential, he described his experience in terms that immediately made me prick up my ears. At first, during the short training session provided by the tour guide, Stefan was dubious about steering what seemed to be a large and quite cumbersome object through crowded streets. But, as the tour progressed, he realized he was weaving his way between people quite naturally—he didn't have to think about braking, speeding up, moving left or right—indeed, his experience was no different from that of simply walking along a crowded street. It was, he said, as if the Segway "were a part of him." And, in a very real sense, it probably was. Our discussion in the previous two chapters of the crucial importance of our bodies to cognition—and the coupling of bodies and brains to the world—leads naturally to this conclusion. As Andy Clark suggests, animals show every appearance of being designed to search out opportunities to integrate worldly resources into themselves, constantly creating new "agent-world circuits" in the process.[2] The radical corollary: not only is cognition extended into the body, as we discussed in the previous chapter, but it can also extend beyond the body and out into the external world.

We have, in fact, already considered this in previous chapters. Let's go back to our cricket robot as an illustration. If you remember, the problem of how to locate a mate is one that was solved by the combination of the design of the cricket's ears (with both an internal and an external route by which sound waves could reach the eardrum) and the way that the pattern of the male cricket's song was attuned to firing of the interneurons linked to the ears. In this way, all the cricket had to do was turn in the direction of the neuron that fired first on any given burst of song, and it would be able to locate the male.

So what at first glance seemed to be a system controlled by internal mechanisms, located in the brain, turned out to depend on a system that was distributed across the body and the environment in important ways: mate finding in the cricket can be considered an example of "nontrivial causal spread" (because the causal links are "spread" across brain, body, and world in a manner that has important, i.e., nontrivial, consequences for how the system works).[3] This clearly links to the ideas of sensorimotor coupling and perception-action loops that we've discussed in the past few chapters, where possessing the right kind of body and exploiting the environment in the right kind of way eliminate the need for any "mirror-like" representations of the world, and where those representations that do exist are heavily action oriented. Clark, and his collaborator, David Chalmers, have simply taken these ideas a step further. In their "extended mind" argument, they put forward the position that, in addition to the kinds of "online" intelligent action described above, external aspects of the environment can also operate as part of "off-line" explicit cognitive processes as well. The crucial element, in both cases, is that something they call the "parity principle" is satisfied. According to that principle,

> If, as we confront some task, a part of the world functions as a process which, were it to go on in the head, we would have no hesitation in accepting as part of the cognitive process, then that part of the world is (for that time) part of the cognitive process.[4]

This argument is an example of functionalism: a philosophical position stating that some entities are best characterized by the roles they play in a process, rather than by their physical form (we have already encountered this in chapter 7 in our discussion of Turing machines where we noted they could made out of any sort of material at all, as long as it could perform the correct algorithm in the correct manner). Clark and Chalmers illustrate their parity principle via a thought experiment involving a chap called Otto, who has a mild form of Alzheimer's disease and, as a result, no biological memory. Instead he owns a notebook filled with all the information he needs to organize his life effectively. When Otto needs to find his way to a certain address, he automatically and unquestioningly retrieves it from his notebook in just the same way that Otto's friend, Inga, automatically and unquestioningly retrieves it from her biological memory. Looked at in terms of its role, rather than its physical properties, Otto's notebook, Clark and Chalmers argue, differs in no way from Inga's biological memory (whether this is, in fact, a good model for memory is something we'll deal with below); it is only habit, convenience, or prejudice that leads us to think that cognitive systems have to be contained entirely by an agent's "skin and skull," and that the notebook doesn't qualify. In essence, the parity principle is another way to point out the false dichotomy between organism and environment that we've been discussing all along. When we take a step back and consider how a cognitive process operates as a whole, we often find that the barrier between what's inside the skin and what's outside is often purely arbitrary, and, once we realize this, it dissolves. We can put it another way: if we think of cognition as an active process, and "mind" as something animals do rather than something they "have," then questions about whether "minds" are things inside the head, or things that can exist outside them, don't really make much sense. The metaphor of "containment"[5] that we are using to think about these things—that a mind is a thing inside a person and distinct from the outside world—begins to break down.

This "causal spread"/"extended mind" viewpoint can help explain why a Segway seems so intuitive and "natural" as a mode of transport, and why people "lose sight of it" as they ride. Take speed control. If the Segway starts going too fast, the platform on which the driver stands begins to tip up very slightly. The driver naturally pulls back on the handlebar as his body begins to tilt away from it, and so the Segway slows down. As it slows, the platform moves down again, and so the handlebar is moved forward. This should sound familiar from chapter 3, and indeed, just as with a Watt governor, a safe speed is maintained by the movements of the platform, the person, and the handlebar dynamically determining and being determined by each other. The person riding the Segway is largely unaware that this is how speed is controlled, and it no doubt adds to the very natural feeling of movement. It is, however, very difficult to argue that the control of a safe and constant speed is down to the Segway driver's explicitly representing an ideal speed and then attempting to control it, or to the driver's body alone controlling speed via an embodied sensorimotor process. Rather, it seems more accurate to say that the process of speed control is distributed across the driver and the Segway. The Segway is more than a tool that the driver is using, because the Segway actively contributes to the process of speed control by exploiting the structure of the driver's body in causally effective ways.

There is, of course, no necessity, nor any suggestion, that the Segway itself is "cognitive" (a criticism that is often applied to this "extended mind" argument and, in particular, the parity principle).[6] After all, no one would argue that a single neuron has any independent cognitive capacities of its own, even though no one would deny that brains as a whole do cognitive work.[7] What the parity principle highlights is the mutuality of animals and their environments, and how these work together to produce adaptive behavior. The environment can be exploited in ways that either help to relieve the cognitive burden in just the same way that making bodies from the right kinds of materials can do so, or (as we'll see later) can even enhance cognitive capacities beyond what the "bare brain" can achieve alone. Stef's Segway adventure nicely illustrates the two important points we're going to explore in this chapter: namely, that the boundaries of the body are, as Andy Clark[8] puts it, "negotiable," and that this is because organisms are always mutually entangled with their environments and do not stand apart from them as some kind of hermetically sealed, disembodied cognizers.

Extending the Schema

> A person is not an originating agent; he is a locus, a point at which many genetic and environmental conditions come together.
>
> —*B. F. Skinner*

Stefan's sense that the Segway was a "part of him" is not at all far-fetched, and has clear links to the work on the body schema that we discussed in the previous chapter. That is, it seems likely that when a Segway becomes "transparent" to a rider, this is because it has been incorporated into the rider's body schema. This is the case for any tool that we use with ease. Get a pencil and poke at something just out of reach. What do you feel? You feel the object that you're poking, rather than just your fingers clutching the pencil (of course, as soon as you read that, you become aware of your fingers, but the primary point of poking is to contact an object, and that's what you do: you contact the object and not the pencil).

Monkeys that have been trained to pull out-of-reach objects toward them using a rake show changes in the specific neural networks that form their body schema, with the result that, in effect, the neural mapping of the monkey's hand becomes elongated to include the tool as well.[9] The activity of the tool-using monkey's brain is recorded from the intraparietal cortex, where bodily and visual stimulation are integrated. In this area, there are so-called bimodal neurons, which respond both to somatosensory information (physical stimulation of the body) and to visual information. So, for example, there are bimodal neurons that respond to stimulation of the hand (the "somatosensory receptive field" or sRF) and also to visual stimuli near the hand (the visual receptive field, or vRF). After the monkeys have had just five minutes of practice raking in food items, the bimodal neurons whose sRF corresponds to the hand show a change in their vRF: it elongates and extends to include the entire length of the tool, so that objects placed within reach of the tool now stimulate the visual neurons, as well as those in reach of the hand. The rake becomes, in effect, part of the monkey's "reaching system." Similarly, there are neurons with sRFs that correspond to the shoulder and neck. Before tool use, their vRFs corresponded to the area that could be reached by the arm alone. Following tool us, these vRFs also expand to include the area that can be reached with the rake.

To produce this effect, the monkeys must actively use the tool to obtain food. Simply holding the rake passively has no influence on the receptive fields of the neuron populations. This shouldn't surprise us, given the emphasis we have placed on active bodies doing things in the world, and also the way in which Freeman's rabbit studies showed that odors needed to be meaningful to a rabbit in order to trigger a response. The incorporation of the tool into the monkey's body schema makes it clear how we can consider the boundaries of the body to be "negotiable" in the way Clark suggests: bodies are not fixed entities from the "point of view" of the body schema, but are continually being constructed and reconstructed in response to the array of resources available in the environment that help simplify the task at hand and/or enhance an animal's ability to complete it. These tools are not simply used by the body but are incorporated into

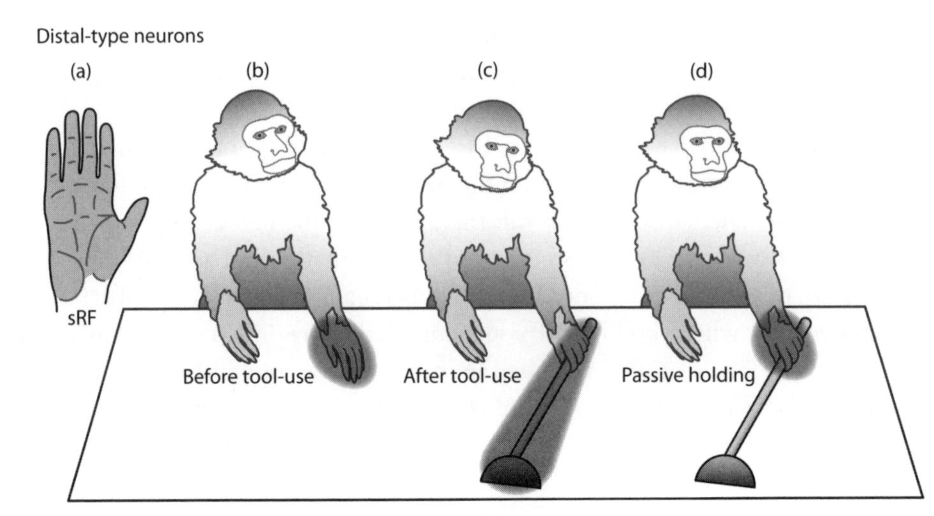

Figure 11.1. If monkeys are allowed to use a rake to obtain a food reward that is out of reach, systematic changes are seen in the receptive fields of two particular kinds of neurons in their brains. Redrawn with permission from Elsevier Publishers.

it. Once we view matters from this perspective, the impossibility of considering an animal independently from its environment becomes clear. It also emphasizes the point that we made earlier that changes in an animal's own body will require a certain level of behavioral plasticity to be maintained; as we have seen, babies have to learn how to use their bodies so that they can become "transparent" equipment, and this requires the negotiation and renegotiation of their bodies over time as their physical dynamics change with growth. The fact that elements of the environment can be incorporated into the body in an equally flexible manner is simply another reflection of this constant negotiation and renegotiation of the body, but this time it is the boundary of the body itself that becomes extended, and new physical dynamics arise as a result of the inclusion of external physical props into the body schema.

These findings automatically counter one of the objections made to the "extended mind" hypothesis and the parity principle: specifically, the objection that external objects are simply sources of input to the cognitive system and not constituents of it. That is, no one would deny that the Segway is coupled to Stef in an interesting causal way, but critics of the extended mind hypothesis argue that, nevertheless, when the pairing navigates through the streets, there is a clear separation between Stef—as the "real" cognitive system—and the Segway, which he is merely using as a tool. If, however, Stef's body schema is altered to incorporate the Segway in a flexible, dynamic fashion (as seems likely), then defining exactly what constitutes the boundaries of "Stef" is more difficult, and it seems more reasonable to conclude that the cognitive system consists of an interesting Stef + Segway hybrid. Only this kind of integration can actually satisfy the parity principle, because the external object is not simply causally linked to the process of steering, navigation, and speed control; rather, by being fully integrated into it, it changes the very nature of the process. Coupling alone isn't enough for us to say that cognition is extended or distributed; it has to be of the right kind.[10]

The Plastic Body

In addition to the raking-monkey studies, there are some very nice scientific studies on humans that make a similar point about this kind of bodily plasticity. In sensory substitution studies, the lack of one sensory modality is compensated for through provision of stimulation via a different, intact sensory system in such a way that it captures important aspects of the missing sense. As this kind of sensory substitution is possible only if there is brain plasticity, the very fact that it works so well is testimony to the flexibility of our brains and bodily systems.

Some of the earliest work in this area, carried out in the 1960s and 1970s, fitted a small "bed of nails"—a grid of blunt metal projections—onto the backs of blind people.[11] The bed of nails was connected to a head-mounted video camera, so it was responsive to visual information (the system is known as tactile-visual sensory substitution or TVSS). Different parts of the grid became active in response to different kinds of input from the camera, and people wearing such a grid would feel this as a tingling sensation on their back. At first, the tingling is all that they feel, but, after a while, people begin to have visual experiences; they experience objects looming up toward them, for example, and are no longer aware of a sensation on their back. As with the tool-using monkeys, the most important factor in generating visual perceptions is that subjects have full control over the camera: they need to control for themselves where the camera is "looking," so that they can learn to correlate their grid-based perceptual experience with the changes they have induced to the input from the camera (which should remind you of Gibson's ideas regarding perception as an active process of information pickup). The most fascinating aspect of these studies, however, is the way that an inwardly directed tactile sensation was transformed into an outwardly directed visual one; people stop experiencing the grid as something poking into their back and start seeing objects outside, in the world. Often, they don't even have the sense that they are engaged in the process of "seeing an object," but are simply aware of the object's being there, with no conscious awareness of the sensations they experience. All of this harks back to Gibson's discussion of the "facial vision" of blind people, which is actually echolocation. TVSS seems to lead to a similar sort of "sensationless perception": the person's perceptual experience is that of seeing, even though it is the tactile system that is being stimulated. Following Gibson, one could then suggest that the nervous system is not "converting" tactile sensations into visual perceptions; rather, this new kind of "hybrid" perceptual system tunes in to the invariants of the environment, via the person's actions with the camera. The experience is visual because the information that can be picked up with a camera will be that of light reflecting from surfaces in the optic array, specifying properties of solid objects and other related aspects of what we consider the visual environment.

The question of whether people really are seeing or just "seeing" is obviously of some interest here. The information about the environment that they detect using the grid on their backs (or, more often these days, from small coin-shaped grids placed on their tongues) is no different from the optical information detected by sighted people. If this is the case, and if the behavior of blind people using this optical information is indistinguishable from that of sighted people, then there is no reason not to call what they do seeing, rather than demoting it to mere "seeing," with heavily emphasized quotation marks. If their perception of the world is the same as that of a sighted person, why not?

The reason is that we tend to feel uncomfortable with the idea that one can see without using one's eyes. This makes sense, because our eyes are part and parcel of our visual perceptual system. But this is precisely where the parity principle comes in: if a grid attached to one's back performs in exactly the same way as the processes that would ordinarily occur in the brain's visual system, why make the distinction? We can reiterate the point we made earlier: it is really just a "within the skin" prejudice to think that only those processes that take place inside our skin and skull are "cognitive," and that external artifacts cannot also be part of the cognitive loop. If we force ourselves to focus on the process as a whole, ignoring any kind of inside-versus-outside distinction, then we are also forced to recognize that the TVSS creates what Clark calls new "agent-world" circuits that provide exactly the same functionality as those that exist in fully sighted individuals.

There are some other fascinating examples of sensory substitution. One, again reported by Andy Clark, is a vest used to help helicopter pilots maintain stability as they fly.[12] It can help ensure that even novices can perform some of the most difficult tasks of helicopter flight, such as maintaining a stationary hover. The suit emits puffs of air onto the pilot's body that correspond to the angle of the helicopter. If the helicopter tilts to the right, the pilot will feel a vibrating sensation on that side of the body. Moving in the opposite direction corrects for these vibrations. The suit can also monitor the pilot's responses to the vibrations (i.e., moving in the opposite direction from the side receiving vibrations) and so control the helicopter. In effect, the helicopter becomes a functional part of the pilot's body via the actions of the suit and, apparently, is so effective that pilots can fly blindfolded. As with TVSS, what matters most is that (a) motor commands affect sensory input—moving the camera influences the tactile input on the back or the tongue, and moving the helicopter controls influences the puffs of air provided by the suit—and (b) these new agent-world circuits are involved in a specific goal-directed behavior. As a result, the suit becomes "transparent equipment," with pilots aware only of the helicopter and its motions, and not the puffs of air or the suit itself. As Clark notes, this is exactly what we should expect from creatures that have evolved to seek out all possible opportunities to exploit reliable properties of the world. As such, it fits beautifully with the ideas we explored in the previous chapters. Plasticity in behavior reflects plasticity in the creation of new body-world circuits—a more "negotiable" body makes for a more adaptable creature.

Objections to Extension

The negotiable nature of the body clearly shows us how agents are tangled up with their environments; a body acts in the world, and it incorporates parts of that world into itself. Is the same true for cognitive processes, however? It is one thing to note that cognitive processes can be augmented and supplemented in various ways, but does this really entail that we have to think of cognitive process as "transcranial"[13] and extending beyond the brain? Some authors make quite an emphatic argument that this is not the case.[14] They argue that only internal brain-based neural processes carry the "mark of the cognitive," and that even those examples that satisfy Clark and Chalmers's parity principle cannot capture these specific features of cognition.[15] One criticism deals with the idea that only internal neural resources have "intrinsic non-derived content." I have to confess I don't fully understand what the authors of this criticism are getting at here—at least partly because intrinsic content is not well defined—but the argument is, roughly, that only internal mental contents (and presumably the neurons and brain tissue by which these are instantiated) possess whatever it takes to be part of a truly cognitive process because of their very nature, whereas Otto's notebook gets its content in a "derived" form (i.e., its contents are in the form of written language, and its meaning is assigned by social convention). This sounds to me like nothing more than a form of "neurochauvinism." Moreover, it is also true that the contents of people's internal mental states and cognitive processes appear to have derived contents. Think of how children learn language: they are taught it socially, and they then use it to transform their cognitive processes in far-reaching ways, so is this derived or nonderived content?[16]

The other major criticism is that there are certain features of (notably human) cognitive processes, like priming and recency effects[17] in memory research, that are not found in their extended counterparts. Otto's notebook isn't vulnerable to these effects in the same way as the memory inside someone's head. But, as Clark has noted, this position argues that, in essence, the "cognitive" should be "marked" by some very idiosyncratic features of how human neural systems apparently work, which is a very anthropocentric, and a heavily neurocentric, view. It seems far more productive—especially from our point of view, concerned as we are with nonhuman cognition as well as that of humans—to adopt the commonsense functionalism of Clark and Chalmers for the simple reason that, if nothing else, it ensures that we stay away from the kind of human-oriented, heavily brain-oriented, view that we have been at pains to dismantle.

One could, however, argue that this seems a bit contradictory. After all, we have also spent a lot of time detailing how the physical structure and design of bodies themselves can be crucial to intelligent action in the world. The "role" played by, for example, the insect eye in compensating for motion parallax depends crucially on its physical properties. Although this is true, it is also true that the Eyebot's eyes, made from materials different from those of a real insect and looking nothing like real compound eyes, achieved the same result in terms of enabling a constant lateral distance to be maintained. In addition, even if the ability to perform some task is crucially dependent on a specific physical substance or design, it doesn't mean that this is true for all the tasks in which an animal engages, nor does it mean that other parts of the same task can't exploit external aspects of the environment effectively. In other words, adopting a functionalist perspective on cognitive processes doesn't force one into denying that the structure of an animal's body matters, or that brains don't make a unique and important contribution to certain kinds of tasks;[18] it simply ensures that we don't construct artificial barriers around those brains and bodies with respect to what counts as a cognitive process. It is also the case that, although the parity principle, as originally proposed, implied that there needed to be a straightforward exchange of like for like—Otto's notebook functioning as internal memory—this needn't be the case.[19] That is, external resources can serve as complements to internal processes, rather than replacements for them, helping to enhance and augment internal processes as well as simply "standing in" for them.

The Mnemonic Mouse (or, Rather, Its Mnemonic House)

> Everybody needs his memories. They keep the wolf of insignificance from the door.
> —*Saul Bellow*

Another reason to take Clark and Chalmers seriously—in particular, the idea of external "memory" as an integral part of a cognitive system—comes from more work in artificial life and robotics, specifically, the creation of an artificial mouse that can engage in delayed reward learning.[20] The ability to learn in this fashion can be tested through the creation of a situation in which a creature must make a decision that potentially leads to a reward—it must decide whether to go left or go right in a maze, say—but the feedback on whether the decision was, in fact, correct is provided not at the time the decision is made, but only later, when a reward is discovered at the end of the maze. As should be clear, delayed reward learning can take place only if an animal can remember the events that led up to its securing the reward. If it can't remember anything, it can't associate the reward with its earlier turning direction.

The interesting thing about the robotic mouse was that it could learn to associate a reward with an earlier cue, but it did so without having any memory for its past choices. None at all. Instead, it had only what was called a "minimal cognitive architecture," in the understated fashion of all good scientists. It was equipped with "whiskers" so that it could detect things by touch; a camera, so that it could see; and a reward sensor that could detect an electronic signal that functioned as a "reward." The sensors corresponding to the mouse's senses of touch and vision were connected to a simple neural network. One set of nodes (which functioned like neurons) in the network represented the state of the sensor (color and intensity of pixels for the camera and touch versus nontouch in the whiskers), and another represented the change in the sensor (differences in color intensity, whether there was a switch from touching to not touching an object). There was also a node corresponding to the reward. In the mouse's motor system, there were neurons corresponding to direction and to change in direction. There was, however, nothing built into this architecture that would allow the mouse to store a memory of its turning decision so that, when it detected (or failed to detect) the reward, it would be able to associate this event with its previous turning decision and so learn the right decision to make. The mouse had no memory. How, then, could it perform so well?

To understand how the mouse achieved its amazing memory feat, we need to know more about the testing environment and how this worked with the mouse's neural network. Let's deal with the mouse first. All the groups of neurons in the mouse's sensory and motor systems were mutually connected to each other. Connections between neurons were then strengthened whenever the neurons were simultaneously active (a simple Hebbian learning mechanism). This should sound familiar because it means that the robot mouse forms time-locked correlations between its different modalities, as we discussed previously for human babies. The other point to note is how this design makes it impossible for the mouse to learn to associate a current event with one that happened in the past (e.g., detecting a stick with its tactile sensor cannot be associated with later detecting the reward with the reward sensor because those events are necessarily displaced in time. The tactile sensor and the reward sensor are never active simultaneously, and so there is no Hebbian learning that can "wire" the association together).

Now, what about the learning environment? This was a T-maze, which is exactly what it sounds like: a very simple maze in the shape of a T. The animal moves along the T until it reaches the junction at the top, and it makes a decision to turn right or left. The reward is placed in one or the other of the T's arms. At the corner of the T-junction, there is an upright stick that can be detected by the mouse's whiskers. This tactile cue accurately indicates where the reward will be: if the stick is on the left, then the reward is in the left arm of the T, and if it is on the right, then the reward is in the right arm of the T. So, to solve the task, the mouse should turn in the direction indicated by the presence of the stick. Of course, when it begins the experiment, the mouse has no "knowledge" of what the tactile stimulus indicates (after all, if this were built into the mouse, it wouldn't have to learn anything). The other relevant aspect of the learning environment is that the inside of the horizontal wall of the T is painted red along its entire length. When the mouse enters the T-maze, it can "see" a red wall at the end of the T, at the point of the junction. The experiment itself is very simple. In each trial, a reward is placed randomly in either the right or the left arm of the T, and the tactile cue is placed accordingly. The mouse with no memory is then given the chance to learn how to find the reward.

So what's the trick? The way the mouse achieves its feat of memory is by "off-loading" it onto the environment. At the start of the experiment, the robot is allowed to run some initial trials in the maze, without the presence of either the tactile cue or the reward. In this way, it learns of the associations between features of the maze and its own actions. When it makes a left turn, for example, it simultaneously activates its right-hand vision sensor because this will detect the red wall, so the mouse learns of the association between that specific motor action (left turn) and a specific visual input (red wall on right).

Now, the reward and the tactile cue are added to the experiment, with the tactile cue always placed so that it indicates the correct location of the reward. The reward sensor is stimulated by the person running the experiment before anything else happens. This simulates the idea that the robot "wants" to find a reward. It has no effect on the robot's behavior because no other associations linked to the reward have yet been learned. The robot is placed in the T-maze and left to its own devices. It heads down the long arm of the T until it gets to the junction. It then makes a random turn—let's say to the left—and heads down one of the T's arms; as it does so, it contacts the tactile cue, and an association forms interconnecting its change in direction (left), the change in visual input (red wall on right), and the change in its tactile sensor (from "no-touch" to "touch"). Let's also assume that, when the robot reaches the end of the left arm, its random decision turns out to have been correct and it gets the reward. Simultaneous activation of the reward signal and the vision sensor (which is stimulated by the red wall to the right) leads to an association between the reward and the red wall to the right. It should now become clear why the red wall is so relevant: although it has nothing whatsoever to do with the task at hand (the wall provides absolutely no cue to the location of the reward), the fact that the robot associates the reward with "red wall on right" means that the information about the robot's previous turn and the tactile cue is indirectly contained in its current situation. A red wall on the right necessarily implies that the previous turn was to the left, and a turn to the left is associated with stimulation of the tactile sensor. The "memory" for the turn and, indirectly, the cue is off-loaded onto the red wall. The delayed learning that occurs is therefore an emergent behavior: there is nothing built into the robot that can allow the correlation between tactile cue and reward to form. It comes about because the red wall functions as the memory for the turn and the cue. In other words, this shows that "memory" doesn't require any kind of dedicated "memory system" built into an animal, but can instead be seen as a property of the cognitive system as a whole.

In the next trial, assuming the same arrangement of cues (stick and reward on the left), the association between "red wall on the right" and the reward will mean that, when the reward sensor is stimulated at the beginning of the trial, the activation produced spreads to the visual areas, which generate the activation pattern associated with the red wall on the right. Now, because "red wall on the right" is also associated with turning left (as this correlation was formed during the initial exploratory activity of the robot in the maze before the reward was added), activation of the vision sensor also spreads to the motor system, and the robot turns to the left. It can then carry on as before and retrieve the reward. As long as the same arrangement of tactile cue and reward occurs twice in a row, the robot will learn the task and will turn in the direction indicated by the tactile cue on each trial (and this will happen inevitably if one runs enough trials: even if the sides are changed randomly, eventually there will be two trials in which the reward occurs on the same side of the maze twice in a row). Now, surely one would say that this feat shows the "mark of the cognitive" (assuming one accepts that learning and memory are cognitive processes), but the key to it is the red wall—a wall that is not itself "cognitive" and is not internal to the robot. An example of the parity principle par excellence.

Thanks for the Memory?

There is an important point to keep in mind here, however. As with the other robot examples that we've discussed, our attribution of "memory" to the mouse clearly reflects our own frame of reference. We can attribute memory to the cognitive system as a whole—the active robot mouse in the T-maze environment that it exploits—but in doing this, we have to be aware that no "memory" as we usually conceive of it exists from the perspective of the robot mouse itself. As Ross Ashby,[21] an early pioneer of cybernetics (the study of complex regulatory systems), pointed out long ago, memory is, in general, an observer-dependent phenomenon. He used the example of observing a dog whimpering as a car goes past. This behavior seems puzzling until its owner tells you that the dog was run over six months before. Now you can attribute the dog's behavior to its memory of being run over. This, suggests Ashby, can lead one astray, because we are then likely to assume that the dog "has" a memory in just the same way that it "has" a patch of black hair; it suggests that we could look for the dog's memory as a real entity that exists inside the dog. But if we consider what we have actually done, we have simply introduced a theoretical construct called "memory" to "fill a gap" in our knowledge. By saying the dog has a memory, all we're really saying is that its behavior can be explained, not by the dog's current state, but by its state six months ago, which we were unable to observe. For Ashby, memory isn't a "real thing" that an animal has or doesn't have; it is just a useful construct that helps us make sense of its behavior.

We aren't used to thinking of memory simply as some kind of theoretical prop. Instead, we think of memory as a very real thing: we are very fond of a "storehouse" metaphor, where information is encoded and stored until retrieved.[22] Various different kinds of memory system have been identified—short-term, working, long-term, episodic, semantic, autobiographical, flashbulb—with distinctive properties suggesting that various different kinds of storage and retrieval takes place. There is also a huge amount of research in cognitive psychology that has investigated the conscious recall of previously learned lists of words. All of this work has been taken to show that memories are particular kinds of stored structures or representations inside our brains, but what we're dealing with is no different from our dog example above.[23] When we ask our experimental subjects to recall a word list, we explain their current behavior by reference to previous events—the presentation and learning of the list—but, just as in the dog example, we haven't actually observed what went on in our subject during the experiment, and we are using the term "memory" to fill in this gap.

What we think is the "mechanism" of memory is really more a redescription of the behavior of the subject. In addition, during the experiment, the subject—the complete agent—is engaged in all sorts of behaviors that aren't considered relevant to the experiment, and so aren't recorded,[24] but which, for all we know, have some bearing on the mechanism by which the second list of words is produced following presentation of the first. One can also look at it from the other perspective, of course, and note that, despite all these other behavioral variables (which will differ across subjects), a consistent set of experimental results is produced. This is what leads to the conclusion that everyone has a memory of the same kind that stores and retrieves information. But it is also the case that the experiment has assumed there is a pure memory function to be discovered, and so many of these experiments beg the question: their design already assumes the existence of the entity that the experiment aims to discover, and so it is not too surprising that a particular kind of memory is found. A storehouse metaphor leads to storehouse experiments, which lead to storehouse memory.

So, just as with experiments on robots and other animals, there is also our frame of reference to consider with respect to experiments on other humans—what looks like a stable structure to an observer from the outside may, from the perspective of the person performing the task, be a more dynamic process of re-creation (or even simply creation). Rolf Pfeifer and Josh Bongard[25] give the example of a fountain: the shape of the water as it sprays out is not stored anywhere as a structure inside the fountain, but results from the interaction of the water pressure and surface tension, the effects of gravity, and the shape and direction of the jets. This gives the fountain structure—not a static, stable structure, though, but one that is continuously created. It is quite possible that memory could have this kind of "structure" and be completely different from our everyday idea of memory.

This is especially likely to be so given that the conscious recall of words is a very specialized human activity. Most of our everyday activities that involve memory are not like this (driving or walking to work; preparing a meal),[26] and it certainly doesn't capture aspects of the daily experience of other animal species. We also learn and memorize many things implicitly—we have no conscious awareness that we have done so, but our behavior changes in ways that reflect our experience—and this kind of implicit memory is undoubtedly common to other animals as well. What we call memory may be much more like the activity of the robot mouse in its maze environment: a process of sensorimotor coordination distributed across animal and environment, in which the animal actively engages,[27] and not simply the storage and retrieval of (explicit) internal representations. One obvious example of this is how, if someone asks me for the telephone number of my lab at the university, I cannot tell her. The only way I can recall the number is to physically dial it, ditto most of the PINs that I use to log on to my email or pay with my debit card. This doesn't seem like the "retrieval" of a stored memory for a particular number; it seems, rather, to support the idea of "memory" as an ongoing process that emerges from my interaction with the environment.

Along similar lines, Paul Broca—the famous French physician after whom Broca's area, a region of the brain associated with language, is named—noted that we have "not a memory of words, but a memory for the movements necessary for articulating words."[28] Even in "traditional" memory research, there is no suggestion that information is simply recalled intact and unchanged on each occasion, as though one were simply pulling a file from a drawer. Rather, the idea is that each recollection alters the memory (because the act of recollection itself becomes part of the original memory). It has also been found that people show different patterns of recall depending on the setting in which it takes place, and on what happened between an event and its recollection. All this suggests that memory may be a highly reconstructive, or even simply constructive,[29] activity that requires action in the world. As we discussed in relation to "concepts" and to the concept of "mind" itself, memory is not a "thing" that an animal either does or doesn't have inside its head,[30] but a property of the whole animal-environment nexus; or, to put it another way, it is the means by which we can coordinate our behavior in ways that make it similar to our past experiences. That is, particular patterns of coordinated sensory stimulation and motor behavior, in conjunction with particular external resources, help to nudge or push us in particular directions, and this is what leads us to construct sequences of behavior that match previous sequences. And so we dial the phone number we need.

With this broader perspective in mind, we can reconsider with intensified interest behaviors like the caching behavior of scrub jays that we discussed in chapter 5. At present, the inference is that these birds have a form of "episodic memory," and there is even some suggestion that they can engage in some form of "mental time-travel."[31] This is, of course, possible (although proving this conclusively is something of an impossibility), but given the findings from the robotic mouse, and the suggestion that our own memory may be more world involving than we suppose, it seems equally likely that scrub jays off-load some aspects of the task onto the environment, and/or that patterns of sensorimotor coordination with the environment make their memory for caching more a matter of online implicit construction than one of explicit recall, or some form of "autonoetic" projection of themselves into the future. Do the birds show changes in their caching postures that reflect particular food types, for example? Do caching and recovery vary depending on the layout of the environment (we know it does when other birds are present) in ways that could indicate the exploitation of environmental structure in useful ways? Similarly, the experimenters use Lego blocks to identify different kinds of caching trays (i.e., to allow the birds to discriminate between a "private" caching tray and a new one), but do these Lego markers may also allow for the off-loading of some other aspects of the task in ways we haven't yet noticed? The answer to all these questions is that we just don't know, because we're using the same storehouse metaphor for animal studies as we use for human studies, and not considering how different kinds of agent-world loops could be formed to support the birds' amazing feats (feats that remain impressive whatever mechanism is being used).

As Pfeifer and Bongard point out, this way of looking at memory seems much more vague conceptually than the "storehouse" metaphor,[32] and indeed it is. This is because we haven't considered it seriously as an alternative, and much more empirical and theoretical work needs to be done to make it more testable and coherent. From our perspective here, this digression into the nature of memory brings us neatly back to the idea of cognitive extension and, more specifically, to the idea that the world itself can be exploited to relieve the cognitive burden placed on the animal and its brain. The red wall in the case of the robot mouse is a case in point. As the red wall is a stable feature of the environment, the memory function can be off-loaded onto it, reducing the need for neural tissue to store any information internally. The red wall functions in exactly the same way as a stored internal representation (like Otto's notebook), so it satisfies the parity principle, and it does so in a cheap and cost-effective way.

Could the Real Cognitive System Please Step Forward?

> In nature we never see anything isolated, but everything in connection with something else which is before it, beside it, under it and over it.
> —*Goethe*

This cost-effectiveness is important because, as we've already noted, evolution is a thrifty process, and organisms that expend more time and energy than do their competitors to achieve the same ends are penalized by natural selection. Andy Clark has dubbed this the "007 Principle" based on the idea that, to be a successful spy à la James Bond, one should know only as much as is needed to get the job done.[33] Knowing too much can cost your life.

In the animal kingdom, Clark shows how several creatures manage to reduce the costs of certain physiological processes by exploiting aspects of the environment. Take the bluefin tuna. Studies of its anatomy and musculature reveal that it is physically incapable of swimming as fast as it does. Studies by fluid dynamicists, however, show that tuna are able to swim faster than their own physical capacities allow because they find naturally occurring currents in the water and then use their tails to create additional vortices, which they can then exploit to gain extra propulsion.[34] As Clark notes, "The real 'swimming machine', therefore, is not the tuna alone, but the tuna in its 'proper context'—the tuna, plus the water, plus the vortices it creates and exploits."[35] Other examples include mole crickets, which construct trumpet-shaped burrows that help amplify the sound of their mating calls, and filter-feeding sponges, which use water currents to reduce the need to actively pump water through their bodies.[36]

What goes for physiology applies equally well to cognition: we shouldn't expect evolved organisms to store or process information in costly ways when they can use the structure of the environment, and their ability to act in it, to bear some of that cognitive load, and save on expensive brain tissue. (We've actually encountered a similar idea in chapter

3, where it was dubbed the "barking dog principle" by Mark Rowlands, in his defense of a Gibsonian approach to perception.) In other words, when we are considering cognitive processes and the forms they might take, we should be careful to "not put into the head what evolution leaves to the world."[37] The concept of the "extended mind" is a way of getting us to see that the "point" of cognition is not to get the world into our heads, but to generate adaptive loops of behavior in the service of action: loops that reflect the soft assembly of neural, bodily, and external resources in dynamic and flexible ways. In this sense, we can again talk about the "complementarity" of internal and external processes, with respect to how they contribute to cognition.

In a distributed or extended approach to cognition, then, actions in the world are not merely indicators of internal cognitive acts, but are cognitive acts in themselves. On this basis, we can make a distinction between "pragmatic acts" that move an individual closer to task completion in the external environment, and what can be called "epistemic acts" that do not (necessarily) aid in the completion of the task itself, but place an individual in a better state in its cognitive environment so that the task becomes easier.[38] Epistemic acts, then, are actions that help improve the speed, accuracy, or robustness of cognitive processes, rather than those that enable someone to make literal progress in a task. If you recall, we considered something similar to this when we discussed ecological psychology in chapter 6, how animals can make information available to themselves by sampling the optic array, and how we move closer to things or reorient them so as perceive the invariants present. From a more complex problem-solving perspective, one very familiar epistemic act example is the way we move Scrabble tiles around to make it easier for us to see the potential words that can be formed. Similarly, experienced players of the Tetris computer game actively rotate each block as it falls from the top of the screen so that they can more easily see where it fits; by contrast, novice players often attempt to rotate the blocks mentally, both increasing their cognitive load and reducing their speed and accuracy.[39]

Another good example, again from Andy Clark, is the manner in which expert bartenders select and array all the glasses they need to fill a drinks order as soon as they get it (indeed, as a distinguished professor and former professional bartender at my own university tells me, this is one of the first things you get taught at bartender school). The distinctively shaped glasses act as cues that aid in the recall of specific orders, and enable speed and accuracy to be maintained in even the noisiest and most distracting environments. Give them a set of uniform glasses with which to work, however, and the performance of expert bartenders falls off dramatically, whereas this kind of manipulation has no effect on novice bartenders: they don't know the trick, and so they attempt to track everything mentally, regardless of the glasses they're given.[40]

My favorite example of how we can "lean" on the world to support and enhance our cognitive processes, however, is something that I bumped into while on holiday in Barcelona. As you may know, Barcelona is famous for its architecture, particularly the work of Antoni Gaudi. His buildings are often described as "warped gothic," but this fails to get across how extremely beautiful and original they are, making fantastic use of the kinds of forms seen in nature. One of the really fabulous things you can see in Barcelona—or at least I thought so—is a replica of the "hanging chain" model he used in the design of the Colonia Güell: an excellent example of how one can use the world as its own best model.

When a string or a cord is suspended from two points, and acted on by its own weight, it spontaneously forms what is known as a "catenary"—a U-shaped curve. Catenaries form the optimal shape for stone arches, because they possess only compression forces and so can stand alone without the need for buttressing. Gaudi exploited this principle in the design for his church by using suspended strings and weights to determine the structure of the arches and how they would need to be built in order to bear these forces of compression. To do this, he first drew an outline of the church on a wooden board (at 1:10 scale) and fixed it to the ceiling. Next, he attached cords to the board, in the positions where the arches were to be placed, and then attached small sacks of lead shot to the catenaries formed by the cords (each sack was one ten-thousandth of the weight that the arch would have to support). Photographs were then taken of the model from various angles, and, when these were turned upside down, the exact form of the structure that would act in pure compression was revealed. In this way, Gaudi could simply "read off" the measurements needed from the model itself, and with much less risk of error than in complex mathematical calculations (where it is always possible to get something wrong). Of course, the drawback is that such models still have to be fairly large and are time-consuming to build. Nevertheless, the hanging model allowed Gaudi to design a building that would otherwise have been impossible for its time, given a lack of computer-aided drawing. Building the model didn't directly help to get the church itself built, but the close interleaving of physical and mental actions in the construction of the model reduced the complexity of designing the church.

Although most of us are not in the habit of designing enormous buildings, we all use the physical world in similar ways to place ourselves into a more cognitively congenial state. Think of the way that we use Post-it notes, Memory Sticks, notebooks, computer files, whiteboards, books, and journals to support our written work, or the way in which we lay out all the ingredients we need for cooking so that what we need comes to hand at the moment we need it, or how we leave our keys right by the door so that we don't forget them on our way out. All of these behaviors reflect a habit of simplifying what would otherwise be cognitively demanding tasks—a habit that is now all-pervasive and underscores just how much of routine human cognition is enacted in the context of environmental supports.[41] In other words, like the paradoxical tuna, there is a true sense in which the real "problem-solving machine" is not the brain alone, but the brain, the body, and the environmental structures that we use to augment, enhance, and support internal cognitive processes.[42]

At the moment, it is not clear the full extent to which other animals engage in these kinds of epistemic acts, nor exactly how they might do so (at least partly because we haven't given it that much attention). What is clear, however, is that, as the philosopher, Kim Sterelny has pointed out, animals certainly do act as "epistemic engineers," altering the world around them so as to change the nature of the informational environment in ways that make it more conducive for themselves or more difficult for others.[43] Many bird and primate species produce so-called contact calls that advertise their location to others, a behavior that clearly simplifies the task of keeping track of other individuals in the environment. The "songs" or long-range calls of whales also fall into this category, and these kinds of vocal dialects and traditions are extremely well developed in cetaceans.[44] Similarly, the suggestion that, by dispersing seeds along particular travel paths, spider and woolly monkeys could help transform the structure of tropical forests, so constructing their own ecological niche—and epistemically engineering their environment to make it easier to locate and remember foraging routes—fits very well with these ideas.[45] Conversely, the use of moss to reduce the conspicuousness of their nests is a means by which birds can engineer their environment to make the cognitive task of their predators that much more tricky.

Recognizing the role that external environmental structures can play in producing adaptive flexible behavior is also useful from a slightly different perspective. As we noted all along, it is all too easy to assume that complex behavior is the result of the operation of some complex internal algorithm, especially when we're dealing with animals whose behavior looks familiar in many ways to our own, like that of many of the primates. Once we understand fully that the physical environment is a resource that animals can use to their advantage and not only an obstacle to be overcome, we can begin to understand that the flexibility and variability of behavior can also reflect how a particular environment affords and impedes certain courses of action. The examples given above don't satisfy the parity principle or the complementarity principle, and aren't instances of "mind extension" in that sense, but they work in a similar way by forcing us to reconsider our default assumption that variable clever behavior must be the result of sophisticated decision making supported by complex, highly cognitive architectures.

Sleeping near the Enemy

A classic case in point concerns the mating behavior of male baboons. Given their large size compared to females, male baboons are able to socially and sexually monopolize adult females during their fertile periods, and the male puts a lot of effort into staying very close to the female, so preventing her from mating with any other males. These close spatial relationships ("consortships") can last from a few hours to a week, depending on the specific population of baboons. Among East African populations, consortships tend to be quite short and are frequently disrupted by aggression from other males, who take over the consort male's position. There are various social tactics that males can employ to either avoid or facilitate a takeover, and these are often held up as the kinds of sophisticated "Machiavellian" behavior that selected for the very large primate brain. The primatologist Shirley Strum and her colleagues (including Edwin Hutchins, a famous champion of the kinds of "extended" cognition described above) took a more situated view of consort behavior, looking at how males acted as complete agents embedded in their environments (rather than as cunning minds plotting and planning in the abstract).[46] In this way, they were able to show that much of the males' behavior was linked to how they were either constrained or afforded certain courses of action by the environment. Their analysis focused on one tactic that they called "sleeping near the enemy," which was a highly successful means by which younger males could displace older ones. Although older males were able to resist consort takeover attempts by younger, more aggressive males during the day, they were less able to do so once the baboons had retired to their sleeping cliffs for the night.

Older, socially experienced males could resist takeover on the plains during the day by using social tactics to divert aggression, such as grabbing a younger animal and using it as a "buffer" against attack. Such tactics require a high degree of visual contact with others, a significant amount of behavioral coordination, and, therefore, sufficient experience with other animals to deploy them successfully. On the sleeping cliffs, however, these kinds of tactics were constrained by the physical affordances of the cliffs. Their height and narrowness affected the males' mobility and their proximity to other animals, and reduced overall visibility. All of these factors detracted from older males' ability to manipulate the situation socially, but they had a positive effect on the more direct, aggressive tactics of younger males. While one could argue that differences in male tactics across the day reflected some form of explicit decision making by males to outwit rivals, their behavior could be accounted for more simply through the recognition that older and younger males had differing repertoires of behavior that reflected their age and experience, and that they were restricted to employing only those behaviors that were afforded by the environment, and were prevented from using others. In the analysis, only the topography of the environment was a good predictor of whether a takeover would occur, and nothing about the males per se. This doesn't mean that baboon behavior is driven solely by the environment (that's as bad as saying it's all driven by what's in their heads). It also doesn't mean that there is nothing about the males themselves that is behaviorally relevant (after all, why is it that they have this particular repertoire of tactics and not some others? And why didn't they develop new tactics for the cliffs to compensate for the differences in habitat?). Nor does it mean that baboons are not, in fact, behaviorally flexible or capable of complex behavior. Broadening the scope of our investigations doesn't mean narrowing our definition of what really counts as smart. Instead, as should now be clear, it means broadening our definition of a "cognitive system," so that we don't assume that only things that happen in brains matter. Or that only animals that happen to have big brains can be considered smart. It's more complex and more interesting than that.

More specifically, it may be that what truly counts as "intelligent" behavior—in the everyday sense in which we usually understand it—is the extent to which animals are able to negotiate the bounds of their body and the world. The more negotiable the body, and the more world-involving the behavioral loops, the more behavioral flexibility there is, then the more "intelligence" we see.

The paradoxical conclusion that we come to, then, is that the difference between humans and other animals may lie in the extent to which we create and exploit the external structures of our world. We augment, enhance, and support our brain-based internal learning processes to an extraordinary degree, hauling ourselves up by our own bootstraps to achieve feats that no one individual could achieve alone. The things that make us smart in our peculiar way now occur outside our heads, as much as in them. And, as we've seen, the same is also true for other "complete agents" in their worlds. The broader perspective we have explored here allows us to see that flexibility and intelligence are properties not of brains alone, but of the embodied, environmentally situated, fully integrated complex that we know more familiarly as an "animal."

Epilogue

In all affairs it's a healthy thing now and then to hang a question on the things you have long taken for granted.
—*Bertrand Russell*

A final word or two. Now that we've explored the ways in which a more "embodied, embedded" approach to the study of cognition and behavior expands and enriches our view of animal life, I hope you'll appreciate why I think an explicitly anthropomorphic approach is unsatisfactory, and worth reconsidering. If body and environment form constituent parts of what we call "mind," it becomes very difficult to see how other animals, with other kinds of bodies, living in other kinds of environments, will "mind" in ways sufficiently like our own to permit the attribution of humanlike mental states. After all, if the ideas of the umwelt, Gibson's ecological theory, and embodied sensorimotor theories have something going for them, then we have to accept that we don't see the "world-as-it-really-is"; we see it only as it reflects our human needs and physical capacities.

The world-as-it-really-is need not, and probably does not, conform to the categories we impose on it, especially given the manner in which language and culture have allowed us to transform our environment to such an unprecedented degree. It seems odd to then imagine that other animals—in their own unwelts, and with their own particular embodied ways of dealing with the world—will slot neatly into our human-defined, linguistically framed categories. Of course, to the extent that we encounter the world in the way that other animals do, then those aspects of our "minding" will be shared. As we've noted more than once, this isn't an either/or argument. It does, however, make the task more difficult because we cannot assume that, when it comes to psychological mechanism, we can use our own peculiarly human conception of an internal, private mind as a guide. The benefit it brings, however, is a wider appreciation of the many ways there are to "be" in the world. By taking seriously the manner in which bodies and the environment help to define what it means to be a cognitive animal, we will gain a more interesting and satisfying perspective on animal psychology, including our own.

NOTES

Chapter One: Removing Ourselves from the Picture

1. http://news.bbc.co.uk/2/hi/7928996.stm.
2. http://www.guardian.co.uk/world/2009/mar/13/journalist-shoe-bush-jail.
3. Osvath (2009).
4. Sample (2009).
5. Macrae (2009). As it turns out, there was quite a lot of journalistic license involved in the reporting of this story, and Osvath was often misquoted (M. Osvath, Pers. Comm.).
6. Sample (2009).
7. Howard (2009).
8. For example, the stone-piling and throwing behavior—and the inferred ability to forward plan—could be a peculiar quirk of Santino's, and not something that is characteristic of all chimpanzees, whether captive or wild. Equally, it might not be forward planning that explains the behavior. Perhaps Santino collects stones when there are no visitors because it is just something that he likes to do—perhaps it is even a kind of stereotypical behavior often seen in captive animals. Then, when visitors appear, and Santino becomes agitated, he finds he has a pile of ammunition at the ready, although that wasn't his reason for piling the stones. Perhaps he does this only when the zoo is open in the summer because rock-piling behavior is only reinforcing at this time of year, when the rocks are warmer to the touch, or perhaps they can't be piled as easily during the winter because they tend to be cold and wet. There is no reason why these explanations should be more likely than the other explanation based on consciousness—and they may not be true, of course—but without more controlled experiments that test whether Santino can learn to plan for other kinds of future events in a similar way, we don't have any evidence-based reason to prefer one explanation over the other. Both can explain why Santino collects stones while calm during the summer and then throws them later when agitated.
9. Gill, V. (2010). Plants 'can think and remember.' July 14. http://www.bbc.co.uk/news/10598926.
10. Jabr, F. (2010). Plants cannot 'think and remember', but there's nothing stupid about them: they're shockingly sophisticated. July 16. http://www.scientificamerican.com/blog/post.cfm?id=plants-cannot-think-and-remember-bu-2010-07-16.
11. Specifically, the scientific article shows that when a single leaf is stimulated by light, there follows a cascade of biochemical reactions across the entire plant, which are

transmitted by particular kinds of cell, known as "bundle-sheath cells." It is suggested that this is similar to the way that nerve impulses are transmitted in animal nervous systems. The "memory" component comes in because the reactions continued for several hours after, even when the plant was placed in the dark, suggesting that the plant was somehow "remembering" the earlier stimulation. As Jabr notes, this is equivalent to saying that, because a pool of water continues to ripple after one has thrown a stone into it, the water is remembering something. http://www.scientificamerican.com/blog/post.cfm?id=plants-cannot-think-and-remember-bu-2010-07-16.

12. Viegas, J. (2010). Human-like brain found in worm. September 2. http://news.discovery.com/animals/worm-human-brain.html.

13. Crudely speaking, if you divided its body lengthways, the two halves would match each other.

14. Tyler (2003).

15. Tyler (2003), p. 274.

16. Tyler (2003).

17. Kennedy (1992).

18. Kennedy (1992).

19. Kennedy (1992).

20. Dennett (1989).

21. Dennett (1989).

22. Dennett (2006).

23. Blumberg and Wasserman (1995) refer to this as the "nominal fallacy."

24. Kennedy (1992).

25. E.g., Heyes (1998).

26. E.g., de Waal (2001), pp. 65–71.

27. De Waal (2001) p. 71.

28. Although, as we shall see later, this isn't the only possible way of characterizing cognition.

29. Blumberg (2007).

30. Gubernick and Alberts (1983).

31. De Waal (2001), p. 71.

32. See, e.g., Byrne and Bates (2006) for this argument.

33. Byrne and Bates (2006).

34. De Waal (1997, 2005).

35. Wilson (2004).

36. Wilson (2004).

37. Hughes et al. (2010).

38. Henzi and Barrett (2003); Henzi and Barrett (2005); Barrett (2009).

39. Inoue and Matsuzawa (2007). For a video clip see here: http://www.youtube.com/watch?v=nTgeLEWr614. Recently, however, Silberberg and Kearns (2009) have shown that this difference could be due to practice. Humans allowed to practice the task to the same extent as the chimpanzee could perform at the same level of accuracy.

40. Morgan (1894).

41. See, e.g., Costall (1993) and Wozniak (1993) on how Morgan's canon has been misinterpreted.

42. Morgan (1890), p. 174.

43. Morgan (1894).

44. Morgan (1894), pp. 54–55.

45. Landauer and Dumais (1997).

Chapter Two: The Anthropomorphic Animal

1. The attribution of human emotions to natural physical phenomena like the weather, mountains, or sea is also known as the "pathetic fallacy."

2. Guthrie (1993).

3. Hume (1889).

4. Guthrie (1993).

5. Guthrie (1993).

6. Guthrie (1993).

7. Neisser (1982).

8. Guthrie (1993).

9. Guthrie (1993).

10. Guthrie (1993).

11. Heider and Simmel (1944). To see the film, try this Web site: http://anthropomorphism.org/psychology2.html.

12. Hashimoto (1966); Morris and Peng (1994).

13. Dasser et al. (1989); Gergeley et al. (1995).

14. Uller (2004).

15. Hauser (1988).

16. Dittrich and Lea (1994).

17. Tremoulet and Feldman (2000).

18. Tremoulet and Feldman (2000).

19. It is also possible that the movements produced by these dots and geometric shapes show animacy and agency because the computer programs used were created by human designers who possess both these attributes. If the patterns of movement are programmed in by a human, perhaps they provide other kinds of cues—ones that have been placed there inadvertently—that people can also pick up and use. Of course, if the program was specifically designed to generate movement trajectories independently of human influence, this wouldn't be a problem. Thanks to Rob Barton for pointing this out.

20. Scholl and Tremoulet (2000).

21. White (1995).

22. Guthrie (1993).

23. Johansson (1973).

24. What we seem to be doing here is picking up on certain "perceptual invariants" that specify a moving human body: see chapter 6 for an explanation of this.

25. Kozlowski and Cutting (1977).
26. Dittrich et al. (1996).
27. Mather and West (1993).
28. Kanwisher et al. (1977).
29. Gauthier et al. (2000); Tarr and Gauthier (2000).
30. Grill-Spector et al. (2004); Xu et al. (2005).
31. Tanaka and Farah (1993); Maurer et al. (2002).
32. Le Grand et al. (2001a, 2001b); Le Grand et al. (2004).
33. Haxby et al. (2001).
34. Goren et al. (1975); Johnson et al. (1991).
35. Simion et al. (2002); Cassia et al. (2004).
36. Brothers (1990); Dunbar (1998).
37. Brothers et al. (1992).
38. Dunbar (1998).
39. Byrne and Whiten (1988).
40. But see Barrett and Henzi (2005) and Barrett et al. (2007) for a slightly different view.
41. Barton (1996, 1999, 2000, 2006).
42. Hence the monkeys and apes have larger brains than the largely nocturnal and insect-eating prosimian species, like bushbabies, lorises, and many of the lemurs.
43. In the last few years, it has become apparent that there is a third pathway—the koniocellular—but this is still not well understood.
44. Brothers (1990); Adolphs (2001).
45. Perrett (1999); Perrett et al. (1982, 1984, 1985, 1987, 1992).
46. Rizzolatti et al. (1996); Gallese et al. (1996).
47. Fadiga et al. (1995); Hari et al. (1998); Cochin et al. (1999); Grèzes et al. (2003).
48. Buccino et al. (2004); Wicker et al. (2003).
49. Gallese (2001, 2005).
50. Gallese (2001, 2005).
51. Whiting and Barton (2003).
52. A part of the brain traditionally associated with aspects of motor control, particularly well-learned "automatic" behavior like riding a bike, or playing the piano proficiently.
53. Vygtotsky (1978); Tomasello (1999).
54. There is, however, an argument that children's ability to attribute and understand mental states in others is an evolved adaptation—a cognitive "module"—that matures at around age four, and it is this that explains our ability to understand others in terms of their private thoughts, desires, and beliefs. Not all agree with the "module" argument, but there is a large, well-established literature that suggests that "theory of mind" (as this capacity is known) emerges quite suddenly around age four (see, e.g., Wellman et al. [2001]), and that some specific form of cognitive maturation is therefore involved. The argument presented above is very much the minority view, but, to my mind at least, it seems to explain more and does so more satisfactorily than does the consensus view. Moreover, there are some strong criticisms of the current "theory of mind" framework

that, among other things, counter the suggestion that theory of mind emerges in a discrete fashion during early childhood: see Reddy and Morris (2004); Costall and Leudar (2004); Leudar and Costall (2004); and Costall et al. (2006).

55. Panksepp (1998).

Chapter Three: Small Brains, Smart Behavior

1. Just in case you aren't familiar with these metaphors, perhaps a brief explanation is in order: the old idea was that there was a "scale of nature," rather like a ladder. The species could be arrayed along each rung of the ladder, with the most "primitive," least intelligent, least "evolved" forms of life at the bottom, moving up to the most advanced, most intelligent. most evolved at the top. Naturally, we humans occupied the top rung. Darwin's theory of evolution by natural selection completely undercuts this idea of a "great chain of being" by replacing it with the metaphor of a tree or a bush (using the metaphor in the same way that we use the term "family tree") to emphasize how all species are related, and that different species' lineages emerge by a branching process, whereby new lineages split from a common ancestor. Each lineage then continues along its own evolutionary path, so that it becomes impossible to rank species along any kind of linear scale.

2. Tibbets (2002).

3. Dyer et al. (2005).

4. Simons (1996).

5. Simons (1996).

6. Holland (2003).

7. Hayward (2001a), p. 616.

8. Grey Walter (1950, 1951, 1953); Holland (2003), p. 353.

9. Holland (2003).

10. Holland (2003).

11. Grey Walter (1953), p. 129.

12. Hayward (2001b).

13. Holland (2003), p. 362, quotes from Grey Walter: "This arrangement is very far from perfect; there is no doubt that, if left to themselves, a majority of creatures would perish by the wayside, their supplies of energy exhausted in the search for significant illumination or in conflict with immovable obstacles or insatiable fellow creatures."

14. Holland (2003) argues that some of Grey Walter's comments on the robots' behavior may have been overinterpreted, and that some patterns may not have been produced in quite the way that Grey Walter suggests. For example, on the basis of his own reconstructions of the robots, Holland (2003) suggests that the "dancing" movements reflect the oscillatory behavior caused by the contact sensor, rather than its response to lights. Holland (2003) makes it very clear, however, that Grey Walter truly was a pioneer of behavior-based robotics: his robots "are still in many ways in the vanguard of modern

robotics, and deserve both recognition and study on that basis" (p. 363). Not too shabby for research conducted using vacuum tubes more than sixty years ago.

15. Grey Walter (1963), pp. 128–129.

16. Quoted in Holland (2003), p. 357.

17. Maris and te Boekhorst (1996); see also Pfeifer and Scheier (1999) for a review of this and other work.

18. This is an area of increasing research interest in animal behavior, specifically in understanding the mechanisms by which individuals in bird flocks and fish schools, and also human crowds, coordinate their behavior. See, e.g., Couzin, and Krause (2003); Dyer et al. (2009); Ballerini et al.(2008).

19. Webb (1995, 1996).

20. Michelsen et al. (1994).

21. A neuron that connects sensory neurons (those that transmit nerve impulses to the central nervous system) to motor neurons (those that transmit nerve impulses away from the central nervous system and stimulate the muscles).

22. Webb (1995, 1996).

23. Lund et al. (1997, 1998).

24. Quinn et al. (2001).

25. Horchler et al. (2004).

26. Hedwig and Webb (2005).

Chapter Four: The Implausible Nature of *Portia*

1. Wanless (1978).
2. Jackson and Wilcox (1990, 1994).
3. Jackson and Wilcox (1994).
4. Jackson and Wilcox (1994).
5. Wilcox et al. (1996).
6. Wilcox et al. (1996).
7. Wilcox et al. (1996).
8. Clark and Jackson (1994).
9. Clark and Jackson (1995).
10. Clark et al. (2000).
11. Harland and Jackson (2000).
12. Harland and Jackson (2004).
13. Byrne and Whiten (1988); Dunbar (1998).
14. Tarsitano and Andrew (1999).
15. Heil (1936).
16. Harland and Jackson (2004).
17. Harland and Jackson (2004).
18. Land (1969a, 1969b).
19. Harland and Jackson (2000).

20. Harland and Jackson (2000); Land and Nilsson (2002).
21. Harland and Jackson (2000).
22. Harland and Jackson (2004).
23. Harland and Jackson (2004); Land (1972).
24. Tarsitano and Jackson (1997); Tarsitano and Andrew (1999).
25. Tarsitano and Jackson (1997); Tarsitano and Andrew (1999).
26. Hill (1979).
27. Hill (1979).
28. Tarsitano and Andrew (1999).
29. Tarsitano (2006).

Chapter Five: When Do You Need a Big Brain?

1. The artist Jennifer Steinkamp's *Dervish* plays on this in a similar, but much less menacing, way. The work consists of four high-definition projections of trees with branches that twirl in a very animal-like fashion. The work is inspired by Sufi priests, who whirl in motion to symbolize the soul's release from earthly ties. The whirling motions of the branches captures this movement, which has all the characteristics of biological motion that we discussed in chapter 2. The result is strange and enchanting—tree as dancing "animal." Unlike triffids, however, Steinkamp's "dervishes" remain firmly rooted.

2. Reader and Laland (2002); Lefebvre et al. (1997); Overington et al. (2009); Schuck-Paim et al. (2008); Sol et al. (2005, 2008).

3. This doesn't mean that animals are in any way superior to plants—that would again be another form of anthropocentric prejudice. Plants are every bit as fascinating as animals and they do some extraordinary things, but this isn't a book about plants, and space doesn't permit an exploration what plants get up to. There are, however, two talks from the famous TED conference—which are freely available online—that do an excellent job of doing so (although Mancuso's talk does verge on the overly anthropomorphic for me): http://www.ted.com/talks/stefano_mancuso_the_roots_of_plant_intelligence.html; and http://www.ted.com/talks/michael_pollan_gives_a_plant_s_eye_view.html.

4. Llinás (1987), p. 341.

5. Hess (1958); Lorenz (1937).

6. For a review, see Bolhuis and Honey (1998).

7. This predisposition emerges during a "sensitive period" around 12–14 hours after birth, and is neither species nor taxon specific: in experiments, a stuffed duck or polecat can work just as well as a stuffed chicken at producing a subsequent preference in chicks. Bolhuis and Honey (1998).

8. Many different kinds of nonspecific experience can trigger the emergence of this predisposition, including handling a chick, exposing it to a tape of the mother's call, or placing it in a running wheel. In other words, the emergence of the predisposition is not dependent on any particular kind of visual experience. Exactly what aspects of the

experience trigger the predisposition is not 100% clear, but handling and wheel running suggest that activity by the chick itself may be crucial, and, presumably, chicks that are played maternal calls will orient toward them, again involving active participation. This would make sense, given the research we'll cover in chapters 10 and 11. Chicks raised in the dark without such an experience do not show any predisposition to approach head-and-neck-shaped objects preferentially. Bolhuis and Honey (1998).

9. Lorenz (1937).

10. At a more "molar" level—that is, dealing with the issue at the broadest possible scale and taking behavior as a whole—we can, of course, say that animals have "instincts" to eat and mate, for example. The point is that these are fulfilled by different behaviors in different ways, depending on the species, and they can involve quite complex processes.

11. Bolhuis and Honey (1998).

12. In order to prevent the unhatched ducklings from being able to hear their own calls, experimenters mute them by painting their syringeal membranes—their "voice boxes"—with surgical glue. The ducklings are then incubated in a soundproof environment so they also can't hear the calls of any other members of their own species.

13. See, for example, Gottleib (1971, 1981, 1991). Unlike the nonspecific experiences required for the emergence of the predisposition in visual imprinting, the auditory predisposition requires a quite specific stimulus: in domestic mallard ducks, unhatched ducklings must be exposed to their own contact-contentment call at an average repetition rate of 4 notes per second if they are to show a preference for their mothers' calls once hatched (mothers call at a rate of 3.7 notes per second). In addition, ducklings call at a much more variable rate (2–6 notes per second) while in the egg than they do once hatched (4–6 notes per second), and only those ducklings that have been exposed to this wide range of notes show a preference for the mother's call following hatching. This exposure also has to occur before hatching; if the ducklings are exposed to this range of calls only after hatching, the predisposition does not emerge (Gottlieb 1985).

14. Porter and Winberg (1999).

15. Macfarlane (1975) .

16. Makin and Porter (1989).

17. Porter et al. (1992).

18. Varendi et al. (1996).

19. Schaal et al. (2000).

20. Varendi et al. (1997). This suggests that babies prefer the familiar smell of the womb over novel smells, but, of course, it also raises the question: why should they this be the case? One reason is that, no matter our age, we all tend to prefer familiar things over unfamiliar things partly because that's how our perceptual systems work: we have greater "perceptual fluency" with things we have encountered many times over. Whether there is any advantage to this per se, or it is simply an unavoidable consequence of how we learn about the world—or indeed some combination of the two—is an interesting question. Alternatively, an inbuilt preference for the smell of amniotic fluid (which would still involve learning, given that the mother's diet seems to influence this) may help prime babies to

learn the smell of their mother's breast and help them to bond with her—much like the process by which the visual predisposition facilitates visual imprinting.

21. Oyama (1985).

22. Mameli (2001); Oyama (1989).

23. Gilbert (2005).

24. Sterelny (2004a).

25. Tebbich et al. (2001).

26. Oyama (1985); Mameli (2001).

27. See Costall (2004) for more on this notion of "mutuality" in relation to the history of psychology in particular.

28. This process is known as "niche construction" (Odling-Smee et al. [2003]). By acting on their environments and changing them in certain ways, organisms can alter the nature of selection pressures that act on them, and so become the selective agents of their own genes. Given this, one cannot view evolution as a one-way process by which organisms adapt to an external environment. Instead it is a two-way process—or, better still, a cyclical process—by which organisms change the world, which changes them, which then changes the world and so on.

29. Pfeifer and Bongard (2007).

30. Dorigo et al. (2000).

31. Pfeifer and Bongard (2007).

32. Pfeifer and Bongard (2007).

33. Although there is evidence that, for example, ants also use visual cues when following pheromone trails, e.g., Baader (1996).

34. Von Uexküll (1957).

35. Reed (1996).

36. Tinbergen (1969).

37. Although it is not entirely clear what "rule" the wasp might be following, it seems likely that we're dealing with a mechanism similar to those seen in the didabots, *Portia* spiders, and ants, where aspects of the wasp's body (and the physics of its perceptual systems) may be involved, or the wasp is sensitive to a reliable aspect of the situation (such as a dead bee: they usually stay exactly where you leave them . . .).

38. Crook (1964).

39. Plotkin (1994).

40. Plotkin (1994).

41. Plotkin (1994).

42. For example, stereo vision is more complex in this respect than monocular vision, so that more brain tissue is needed to help guide the behavior of an animal with binocular vision compared to one without.

43. Dawkins (1976); Dennett (1984); Plotkin (1998).

44. Dawkins (1976).

45. It some senses, it is perhaps slightly misleading to state that brains allow for more flexibility and learning in an absolute sense; after all, the whole point we've been trying to make here is that flexibility and "intelligence" are relative to an animal's needs. A wasp

is not as behaviorally flexible as a baboon, it's true, but, then again, it doesn't live in a baboon's world, so that kind of one-to-one direct comparison is slightly misplaced.

46. McComb et al. (2001).

47. McComb et al. (2001).

48. McComb et al. (2001).

49. See Clayton et al. (2007) for a good overview of all this work.

50. In all cases, the experimenters removed the food before the birds were allowed to recover their caches so they could be sure that the birds were not just using their sense of smell to identify the foods in each site. The cache sites are ice-cube trays filled with sand, so it was a simple matter to recover the food and provide the birds with the identical, but food-empty, tray.

51. Clayton and Dickinson (1998).

52. Clayton et al. (2007).

53. Dally et al. (2004, 2005).

54. Emery and Clayton (2001). The proximate mechanisms by which scrub jays achieve all these feats are still open to debate. While some argue that they rest on high-level capacities, such as the ability to see the situation from another bird's point of view (Emery and Clayton [2001]; Clayton et al. [2007]), it is also possible that simple "rules of thumb" can explain such findings. For example, the birds may have an implicit rule: "If another bird is present, cache where it is darker." The birds don't have to understand why they have this rule, or even know that they have it all. As long as the presence of another bird alters their caching strategy in this kind of systematic way, the system will work. Similarly, the fact that birds recache only if they themselves have stolen from another bird's cache doesn't necessarily mean that they are thinking, "I'd better recache my food because if that were me, I'd go in and steal it." Rather, the tendency to recache may be dependent on experience in much the same way that imprinting is dependent on experience. In other words, seeing other birds cache and then stealing from those caches may simply increase the tendency to recache one's own food in a correlated, but not causal, fashion. The birds don't need to explicitly recognize or be aware of the fact that their caches are vulnerable, or that they should move them around so that other birds can't steal them. If birds are in an environment where lots of competitors are present, and this increases the likelihood that they will have the opportunity to steal from other birds' caches, this may also trigger higher levels of recaching of their own food when other birds are present. There doesn't have to be any causal understanding on the part of the birds for these two behaviors to co-occur. Both processes can be simply mediated by the presence of more competitors, more caches, and more opportunities to pilfer. As we mentioned earlier in relation to Santino the chimpanzee, just because we can think up an alternative doesn't mean that this explanation is correct and the high-level explanation is wrong. It just means that, as before, the available data are consistent with both the "high-level" and the "low-level" explanations, and so we have no means of distinguishing between the two. Different kinds of test are needed to fully tease these apart.

55. Churchland (1986), p. 16.

Chapter Six: The Ecology of Psychology

1. See Noë (2008), who uses this dancing metaphor to great effect to describe his work on consciousness. Interview on Edge Web site. www.edge.org/3rd_culture/noe08/noe08_index.html.

2. Noë (2008). Interview on Edge Web site. www.edge.org/3rd_culture/noe08/noe08_index.html.

3. So, just to be completely clear, this chapter is making no claims about the conscious experiences, or otherwise, of other animals. It is just taking Noë's analogy as applicable to all psychological processes, whether conscious or not.

4. You may also see it referred to as the theory of "direct perception," or, as Gibson called it, a theory of "information-based perception" (Gibson [1966], p. 266).

5. Gibson (1979); Reed (1996).

6. Gibson (1966), p. 1.

7. Gibson (1966), p. 1.

8. Gibson (1950).

9. Dewey made this point in a classic paper criticizing the notion of the "reflex arc"—the idea that a sensory stimulus comes in and a motor response goes out in a strictly linear fashion. This linear process fails to capture the active exploratory nature of how animals encounter their environments, and how sensory and motor processes operate together in a mutually reinforcing cycle—i.e., the process of "sensorimotor coordination"—where actions take place over time, and these constantly change the nature of the stimulus information available, which then changes the possibilities for action, and so on and so forth. Instead, the classic view promotes a kind of "dualism," where a stimulus is seen as a completely distinct entity that produces a similarly distinct and independent response. As Dewey notes, it makes more sense to treat stimulus and response "not as separate and complete entities in themselves, but as divisions of labor, functioning factors, within the single concrete whole" (Dewey [1896]); see also Rockwell (2005); Reed (1996).

10. Gibson (1966), p. 5.

11. Gibson (1979). As the epigraph at the beginning of this section reveals, Gibson used the word "complementarity" to get this idea across. This is an interesting choice of word, for it was originally used in physics to refer to phenomena like light that can act as both a particle and a wave ("wave-particle duality"). Assuming that this is what Gibson had in mind when he used the term, it suggests he was attempting to get at the way animals and their environments are mutually dependent on each other, and so are "co-defined" —that is, the term "animal" naturally implies the existence of an environment in which the animal lives, and the term "environment" naturally implies there is an animal to live in it (Wells [2002]). In other words, the term "animal" is really just one way of referring to the animal-environment duality, and "environment" is the other, just as one refers to waves or particles depending on the specific aspect of a physical system one is investigating. This doesn't mean that there is no "boundary" around the organism, as

some take this to suggest. Clearly, our ability to differentiate animals from their environments shows this to be false. This doesn't detract from the fact that animals in the natural world are never isolated from their environments but remain "embedded" in them, and in a mutually interdependent relationship with them, at all times.

12. Michaels and Carello (1981).

13. We'll elaborate on this more "embodied" way of thinking in chapters 9 and 10.

14. Gibson (1979), p. 129.

15. There have been many debates about whether "affordances" are objective animal-independent or subjective qualities. I'm not going to go into all this, because it's rather tangential to the main point we're making here, which is that most animals do not see "objects" in the detached way that we see them—which we achieve via culturally supported language skills—but simply as opportunities for action in the world.

16. See Wells (2002), who uses this example to great effect.

17. Powers (1973); Cziko (2000).

18. Cisek (1999).

19. Bourbon (1995).

20. Bourbon (1995).

21. Bourbon (1995).

22. E.g., Marr (1982). In what follows, the criticisms of the conventional view concern only its conceptual foundations; there is a lot of brilliant empirical work that continues to stand, regardless of the interpretation placed on it.

23. Hyman (1989).

24. Hyman (1989).

25. Helmholtz (1868).

26. As Hacker (1995) notes, "this literally makes no sense. An electrical impulse can no more become a sensation than a fact can become an event."

27. Gregory (1980).

28. Blakemore (1977), p. 91. As Hacker (1991) puts it, viewed in this way, "the central nervous system is beginning to look more like the Central Intelligence Agency" (p. 303).

29. Marr (1982).

30. Noë (2009). An even more extreme version of this view is that we don't even have the experience of the world that we think we do, and that the environment we see is some kind of "grand illusion"—for more on this, see Noë (2009).

31. Noë (2009) gives more details of the other "problems" associated with the impoverished stimulus we receive—the uneven resolving power of the eye, the instability of the retinal image, the blind spot—and how these are used to argue for the "creation" of our environment inside our heads.

32. As it says in a leading textbook on the subject, "the appearance of our perceptions as direct and precise images of the world is an illusion." Kandel et al. (1995), p. 368.

33. Bennett and Hacker (2003).

34. Although if "you" are your brain, to whom is the brain telling all this stuff? Hmmmm.

35. This is also known as the "homunculus fallacy": one that Descartes warned against when developing his theory of visual perception, and partly why he played down the significance of the upside-down image that so puzzled Kepler. That is, to say that all we "see" is the upside-down image on our retina would require someone else to be sitting in the brain and doing the seeing for us.

The artist Tim Hawkinson has a lovely piece, *Untitled: Ear/Baby*, that plays on this idea. It shows a drawing of an ear, which hangs suspended from a wall at an angle. As you move around to see behind the drawing, a sculpture of the inner ear extending backward from the drawing is revealed, and close inspection shows it to be a small fetuslike creature, as opposed to the cochlea that one expects. As my friend John says, "It looks like the long sought 'homunculini' that hears in the ear."

36. Hacker (1991), p. 305.

37. Blakemore (1977), p. 66.

38. Hacker (1991), p. 291.

39. Noë (2009).

40. Hacker (1991).

41. So this is a different sense of the word "information" from the one with which we're all generally familiar, where information is something that is transmitted or communicated, and then requires processing. In Gibson's use of the term, information is just "there."

42. Gibson (1979), p. 14, says that his view "is wholly inconsistent with dualism in any form, either mind-matter dualism or mind-body dualism. The awareness of the world and of one's complementary relations to the world are not separable."

43. This means there is a distinction to be made between sources of stimuli and stimuli themselves: the former are objects, events, places, and surfaces; the latter are the patterns and transformations of energy at the animal's receptors that reflect those sources.

44. Gibson (1979).

45. Gibson (1979).

46. Gibson (1979).

47. Noë's (2004) sensorimotor theory of perception is similar to Gibson's ecological theory in that both argue that sensorimotor action in the world is the means by which invariants are detected and so perceived. It differs in that, while Gibson's theory argues that transformations of the optic array are not dependent on specific kinds of motor action (i.e., different forms of motor action can transform the array in similar ways and so specify the same invariant), the sensorimotor hypothesis argues that the detection of a specific invariant requires a specific type of motor action, and so the motor action becomes part and parcel of the perception. Put more formally, motor actions are constitutive of perception in the sensorimotor theory, but are only causally important to perception in Gibson's theory. For more on this, see Mossio and Taraborelli (2008).

48. For example, right about now people often say: "What about visual illusions? Don't they show that the world we see is not 'real' but only constructed in our heads, and therefore depends on representation?" This is certainly the argument that Richard

Gregory uses, for example. Gibson, however, argues that illusions are the result of inadequate information available in the environment or deficiencies of the perceptual process that prevent pickup. Many laboratory experiments, for example, deliberately limit the ability to pick up information by imposing a very precise stimulus and preventing subjects from actively exploring their environments. Similarly, the fact that in the Muller-Lyer illusion, for example, people see the two lines as different in length, when they are, in fact, equal, does not mean that the illusion is an entirely subjective brain-created phenomena. It means merely that the lines themselves are not the stimulus; it is the lines in combination with the other lines that provide stimulus information, and these that generate the perception. As Gibson notes, it would require a very special kind of selective attention to see the line segments in isolation from the other parts of the drawing (Gibson 1966). Similarly, Noë (2009) makes the point that "our perceptual skills have evolved for life on earth, not life in an environment where objects materialize and vanish at the whim of supernatural deceivers. . . . So the fact that we are vulnerable to deception—in the psychology lab or at the movies—just reveals the context bound performance limitations of our cognitive powers. It does not show that our cognitive powers are radically deluded!" (p. 142).

49. Gibson (1979).

50. Gibson (1979), p. 254.

51. Gibson (1970), p. 426.

52. Gibson (1974), p. 42.

53. For a comprehensive array of papers supporting the ecological approach. see http://ione.psy.uconn.edu/publications.html.

54. Gibson (1966), p. 27.

55. Gibson (1966).

56. Reed (1996).

57. Rowlands (2003).

58. Good overviews are to be found in Gibson (1966, 1979); Reed (1996, 1996); Michaels and Carello (1981).

59. This isn't to say that ecological psychology is a perfect theory and has got it all right—after all, what theory can do that? There are points on which Gibson is inconsistent about his own arguments (the notion of affordances as animal independent being a case in point). It does, however, present a genuine alternative to the conventional view, particularly if one is evolutionarily minded, given its inherent "thriftiness"; and, as mentioned earlier, it is also a theory that has a good deal of empirical support.

60. Reed (1996).

Chapter Seven: Metaphorical Mind Fields

1. See, e.g., Boroditsky (2000).

2. Lakoff and Johnson (1999).

3. Lakoff and Núñez (2001).

4. Holyoak and Thagard (1996).

5. Hesse (1966); Ortony (1979); Leary (1990).

6. There are two schools of thought with respect to how metaphor is then used following this catachresis. One is that, as a field develops, and we gain more knowledge and understanding, so the need for metaphorical usage falls away. Another view is that metaphor always forms the core of science, and that metaphorical usage does not—and indeed cannot—give way to literal understanding. Instead, the original metaphorical usage simply is reified (creating the impression of literal usage) or replaced by a new metaphor that provides greater potential for scientific exploration. In this view, our scientific understanding inevitably remains fundamentally metaphorical, even as it becomes more precise in its explanations and accurate in its predictions. Note that this does not mean that our understanding of the world is ungrounded. To accept that all scientific knowledge is metaphorical is not to deny that the world is amenable to accurate prediction and explanation, via application of the scientific method.

7. Tooby and Cosmides (2005), for example, state: "The brain's evolved function is to extract information from the environment and use that information to generate behavior and regulate physiology. Hence, the brain is not just like a computer. It is a computer—that is, a physical system that was designed to process information. . . . The brain was designed by natural selection to *be* a computer" (p. 16, emphasis in the original).

8. Draaisma (2000).

9. Draaisma (2000).

10. The "neobehaviorists" Edward Tolman and Clark Hull both made use of "intervening variables" in their studies, for example. See Malone (2009) for a good review of differing schools of behaviorist thought.

11. Early behaviorists, like John Watson, argued that only behavior should be the subject matter of psychology because internal processes (like "mind" or "consciousness") could not be measured or observed in any way. There is some debate over Watson's views in this respect. One argument is that, in rejecting "mind," Watson was implicitly accepting the view of the mind promoted by Descartes (that there is a subjective "mental" world that stands in contrast to the external world of objects); after all, to say it's there, but we can't study it, is to accept that such a thing exists (e.g., Costall 2004). The other argument is that Watson was actually rejecting this Cartesian concept of mind as subjective and inaccessible, and was instead going back to Aristotle, by treating mind as a form of "activity," as behavior, and was suggesting that, once we have accounted for all of an organism's activities, there will be nothing left over that we can call "mind" or "consciousness." In this sense, "mind" is simply the activity of the living body (Malone [2009]). You can make up your own "mind" about which of these views is correct . . .

Radical behaviorism—the brainchild of B. F. Skinner, perhaps the most influential psychologist of the twentieth century—took a different approach. Skinner expressly denied the mind-body dualism inherent in those forms of behaviorism that excluded mind on the basis of its unobservability (which Skinner termed "methodological behaviorism"). By contrast, radical behaviorism simply doesn't distinguish between an inner, subjective world and an outer, objective one. As Skinner (1987) writes, "cognitive psychologists like to say 'the mind is what the brain does' but surely the body plays a part? The mind is what the body does. It is what the person does. In other words, it is behavior" (Skinner [1987], p. 784). There are some clear links between methodological behaviorism and the computational approach of modern cognitive psychology (odd as that may seem), whereas Skinner's radical behaviorism has much more in common with Gibson's ecological psychology (especially the rejection of "inner" and "outer" worlds), both of which reflect the "mutualistic" view of the organism-environment relationship: see Costall (2004).

12. Miller (2003). Among other things, learned sickness aversion, where the stimulus that causes sickness is displaced in time from its consequences, seemed to suggest that more was going on inside an animal's nervous system than the simple linking of stimuli and response in a classical (Pavlovian) conditioning model. In addition, it is clear that animals are prepared evolutionarily to learn some things more easily than others; pigeons can be taught to peck a key to receive food, but cannot learn the same response in order to avoid an electric shock. This said, it is also clear that radical behaviorism was misunderstood: the rejection of a division between internal, private events and external public events was often misconstrued as the denial that any such thing as a "mind" existed, whereas Skinner's argument was that "mind" wasn't a "thing" inside us, but part and parcel of what we get up to in the world; we call this "mind" because we have been taught to do so over the course of our development in our particular social environment.

13. E.g., Pinker (2003).

14. Given this, the precise mechanics of how the tape head reads and writes don't matter; it's just taken for granted that this is achieved in some effective way.

15. E.g., Newell and Simon (1974).

16. Some people, most notably the philosopher John Searle, do not hold with the computational/information processing metaphor and the notion of "multiple realizability"—the idea that a brain can be made of anything that can achieve the same functions as a biological brain. Searle maintains that there is something causally and constitutionally important about a brain made from neurons. This doesn't mean that neurons are somehow "magical"; it means only that the material constituents of biological brains should not be considered as irrelevant to what brains are capable of doing. Whether brains do what they do because of how they're made remains an empirical issue, but, as we'll see, it is clear that other aspects of body structure are crucial to the production of adaptive behavior, so there is no reason in principle why the same shouldn't be true of our brains and nervous systems. Conversely, given that we have no evidence or solid empirical grounds for insisting that only a biological brain can do what it does simply because it is biological, we shouldn't be too hasty to dismiss the idea. After all, we have artificial hip joints and cochleas that work almost as well as the real thing. The short story "They're Made out of Meat," by Terry Bisson, makes this point beautifully: two aliens engage in a conversation about life on Earth, during the course of which one of

them discovers, with mounting incredulity, that humans are able to think, dream, love, and even have conscious experiences with brains that are "made out of meat."

The real problem with both sides of this argument, of course, is that they adhere to the idea that it is the brain alone that does all the thinking, feeling, and loving. The point we've been making so far is that these are things that only whole organisms can do.

17. Fodor (1999).

18. Block (1995). This position is known as "functionalism." To put it in everyday terms, what matters is that the job gets done, not exactly how this is achieved in precise physical or mechanical terms.

19. There are other forms of computational models, based on "neural networks," that do not involve the explicit manipulation of symbols according to a set of rules in an algorithmic fashion, and so they don't fall prey to these kinds of criticism. Although it is true that these alternative models are based explicitly on the neurons of the brain, it remains important to note that (a) they are also very abstract models; neural network models do not resemble real brains in any substantive way, and (b) a lot of network models conform to the same "input-cognition-output" structure as the classic symbolic approaches: they have "input layers," "hidden layers," and "output layers." The main difference in the two approaches lies in the fact that the "representations" in a neural network are distributed across all the units and are therefore not made explicit as they are in symbolic models. This "subsymbolic" approach makes them more attractive in many ways because they can capture context in a fashion that symbolic approaches can't, for example, and they can also give rise to "emergent" properties that were not explicitly programmed by their designers. In these and other ways, connectionist models provide a clear alternative to symbolic computation. Neither approach, however, incorporates the role of the environment, or the body, in the manner that Gibson or Noë argues for. Of course, classic symbolic approaches and neural networks are both perfectly appropriate as models of cognition under particular circumstances; it's when their metaphorical nature gets forgotten that problems arise.

20. Wells (2006).

21. Wells (2006).

22. Turing made it a one-dimensional tape rather than a two-dimensional sheet of paper, as this makes the construction of the abstract machine easier, and he argued that anything you could do on a sheet of squared paper you could do on a one-dimensional tape.

23. Wells (2006).

24. Wells (2006).

25. Indeed, Wells (2006) suggests that the inclusion of human "states of mind" in his first models was probably viewed by Turing as a weakness in his approach. Accordingly, Turing went further in his subsequent analyses and introduced another concept: a "note of instruction." This was a physical manifestation of the state of mind of the human computer at a particular stage in the calculation, designed to enable her to take a break from the sum being computed, and then to pick up from where she left off. A note of instruction was simply a way of identifying the stage that had been reached in a computation and

what needed to be done next. To cut to the chase, the inclusion of such a note in Turing's models meant that the human mind no longer needed to be an integral part of the machine. Instead, as Wells puts it, the human computer changed roles from that of an active participant to that of a mere interpreter of the note of instruction. From Turing's mathematical and wholly practical perspective, this was a great advantage because it meant that, in principle, the human computer could be done away with altogether, and the process of computing a number could be achieved entirely mechanically by a form of mechanical "interpreter" that performed the same function. One can also see how the removal of the human mind from the workings of the machine would help create the impression that the Turing machine was a model of the mind that sat entirely in the head.

26. Wells (2006).

27. Wells (2002).

28. Wells (2002, 2006) also shows how Turing's theory can capture "effectivities," which are the capacities and abilities of animals that allow them to act on affordances. We haven't dealt with these here, mostly because of space limitations and because they aren't essential to an understanding of the fundamentals of ecological psychology. Moreover, I'm not sure they are needed; as we saw in the last chapter, the concept of an affordance as defined by Gibson fully incorporated the abilities of animals in a way that was intended to deny the dualism of organism and environment. The inclusion of a term that specifies the animal alone seems to defeat the object. I could be completely wrong, of course.

29. Wells (2002).

30. Wells (2002).

31. Haugeland (1995).

32. Dreyfus (1992); Brooks (1999); Pfeifer and Bongard (2007).

33. Wells (2002).

34. A point made very cogently and clearly in the various papers by Rodney Brooks (1999).

35. See Byrne and Bates (2006), and see Barrett (in press) for a critique.

36. E.g., see Penn et al. (2008).

37. As an aside, it is also the case that this linear stimulus-response-based model is drawn directly from behaviorism. The "cognitive revolution" simply changed an S-R psychology into an S-X-R psychology, or, more accurately, into an Input-Processing-Output model. As both Cziko (2000) and Costall (2004) argue, the cognitive revolution is, in essence, just a quibble about how much goes on in the head of the animal—and whether it is amenable to scientific study—and not the wholesale rethinking of the relationship between organism and environment in the manner of Gibson's ecological theory or PCT.

38. Van Gelder (1995).

39. Van Gelder (1995). This is true whether we are considering classic symbol-processing models or neural network models.

40. Van Gelder (1995).

41. The particular school of Evolutionary Psychology, promoted by Leda Cosmides, John Tooby, Steven Pinker, and David Buss, is entirely predicated on the computational

theory of mind, which it takes to be axiomatic. Hypotheses are tested on the assumption that the brain is actually (and not simply metaphorically) a computational device, and that cognition is information processing, quite literally. Pinker (2003) probably makes the strongest claim for this approach. See Wallace (2010) for a critique of Evolutionary Psychology's commitment to the computational theory of mind.

42. Wheeler (2005).

43. Wheeler (2005).

44. Wheeler (2005).

45. Wheeler (2005).

46. Clark (2008).

47. Clark (2008).

48. Wheeler (2005).

Chapter Eight: There Is No Such Thing as a Naked Brain

1. Freeman (1991, 1995, 2000); Freeman and Skarda (1990); Skarda and Freeman (1990). See also Kelso (1995) and Spivey (2007) for dynamical approaches to human cognition.

2. Pfeifer and Bongard (2007).

3. Low-energy states that are, as a result, stable.

4. Pfeifer and Bongard (2007).

5. A train station, with people moving around, getting on and off trains, and finding their connections, is chaotic in just this technical sense. People rushing for the station doors in a panic because someone shouted "fire" is chaotic in our everyday sense of the word, but not in the mathematical sense.

6. Pfeifer and Bongard (2007).

7. Cisek (1999).

8. See also Cziko (2000).

9. Hebb (1949).

10. Freeman (2000); Dreyfus (2007).

11. Dreyfus (2007).

12. See, e.g., Shepard (1984). This also makes the point, as we noted in chapter 5, that Gibson's theory doesn't exclude internal processes per se, only the formation of static internal representations that compensate for the lack of information provided by the senses. As we noted, Gibson's theory specifically includes the nervous system as part and parcel of an animal's perceptual system, and his views seem to be borne out by Freeman's work.

13. Freeman (2000).

14. Pickering (2010).

15. Dreyfus (2007).

16. Possibly this was due to lack of oxygen to the brain, and impending thrombosis, but I prefer to think that Dreyfus's explanation is the correct one.

17. Clark (1997).

18. In his recent book, however, Anthony Chemero (2010) makes a case for why we can, indeed, do without representations, drawing on both dynamical systems theory and ecological psychology much as we do here, but with some important differences. This book came out too late to be discussed in detail here—hence a brief mention in the footnotes—and the argument is more philosophically oriented than ours, but it is definitely worth reading for another take on these same issues.

19. Specifically, we'll deal with the take on these experiments provided by Maturana and Varela (1998).

20. Sperry (1945, 1951).

21. Maturana and Varela (1998), pp.125–126. See also Johnson and Rohrer (2007)

22. Maturana and Varela (1998); Johnson and Rohrer (2007). Recall also that we made a similar argument in our discussion of PCT: a gymnast's perception of her behavior differs from the judges' assessment of it.

23. Johnson and Rohrer (2007).

24. This also predicts, given that the structure of an animal's nervous system plays a crucial role in mediating the ways in which it encounters the world that, as animals require more flexibility and greater independence from environmental contingency, we should expect to see greater elaboration of their nervous systems. This is, at least partly, achieved through the building in of some "redundancy." Put simply, what this means is that an animal's sensory and neural apparatus is elaborated so that it has more than one way to sense and process the world—it can hear, smell, and touch it as well as see it, for example—so that it can acquire information about what's going on in the world in many different ways. This is like having two pilots fly a plane: if one should drop dead of a heart attack, the other can take over and ensure the plane lands safely. Similarly, if one particular sensory modality fails, the animal can still keep going, as long as the remainder continue to work effectively. Unlike the two pilots, however, the different modalities are not fully redundant (they don't do exactly the same thing) but only partially, as this gives even greater flexibility. We explore this in more detail in the next chapter.

25. Heidegger (1927).

26. Eastern philosophers, by the way, have argued this all along: see Varela et al. (1991) for a nice review.

27. For a more detailed, but very clear and accessible, explanation of this, see Wheeler (2005).

28. This is certainly how Vygotsky would see it, I think.

Chapter Nine: World in Action

1. Clark (1997).
2. Clark (1997).
3. Mataric (1990).

4. It would, more technically, be called an "egocentric" map or action plan, centered on the rat's body. An allocentric map is one that is centered on the external environment.

5. Clark (1997), p. 49.

6. Mataric (1990).

7. Thach et al. (1992).

8. Granzow (2010).

9. This raises interesting questions about what it means to say that this is an "illusion." This is partly due to the way we choose to describe the phenomenon. That is, our perception is understood as an "illusion" when it is based on a spectral description, where pitch is thought to arise from the low component in the spectrum, or the fundamental frequency. When this frequency is absent, and the pitch associated with it is still heard, it is called illusory or phantom. However, if we instead use temporal descriptive frameworks, the term "illusion" becomes a misnomer, because the fundamental corresponds to the period, or overall rate of vibration of the waveform, and this does not depend on the presence of the fundamental frequency component (John Granzow, pers. comm.).

10. Seither-Preisler et al. (2007).

11. Brooks (2002).

12. Brooks (2002).

13. Brooks (2002), p. 30.

14. Brooks (1999). This approach initially went down like the proverbial lead balloon in an academic community firmly committed to the computational-representational approach. The paper in which he first reported on his robotics work was unanimously rejected for publication and took several years to see the light of day in a journal, even though today it is an acknowledged classic in the field.

15. Brooks (1999).

16. As Bourbon (1995) notes, although these robots act without plans or representations, they are still "linear" "stimulus-response" devices that control action rather than perception, and so they cannot produce an unvarying response, given a highly variable environment, in quite the same way. That is, they can operate effectively in the world without the need for any internal "cognitive" processing, but without the same level of flexibility as a PCT system.

17. Brooks (1999).

18. Pfeifer and Bongard (2007).

19. Cruse (1990).

20. As we've mentioned before, in the *Portia* spider example, and those in this chapter, there is no need to imagine that the rule exists in the organism's head in this form. This is simply how we see it from the outside.

21. See Alberts (2007) for a review of this work.

22. Schank and Alberts (1997); Schank and Alberts (2000).

23. May et al. (2006).

24. May et al. (2006).

25. McGeer (1990).

26. For a video of one of the most state-of-the-art passive walkers in action, see http://ruina.tam.cornell.edu/hplab/downloads/movies/Steve_angle.mov.

27. Collins et al. (2001).

28. Collins et al. (2005). These findings suggest that human steady-state walking may well be more dependent on our anatomy than on motor control of the nervous system.

29. Pfeifer and Bongard (2007).

30. Pfeifer and Bongard (2007).

31. Clark (2001).

32. Yokoi et al. (2004).

33. Killeen and Glenberg (2010).

34. Searle (1983), p. 230.

35. Pfeifer and Bongard (2007).

36. Pfeifer and Bongard (2007).

37. Lichtensteiger and Salomon (2000).

38. Pfeifer and Bongard (2007); Lichtensteiger and Salomon (2000).

39. Iida and Pfeifer (2004).

40. Iida and Pfeifer (2004).

41. Clark (1997, 2008).

42. Clark (1997); Thelen and Smith (1994).

43. This is a variant of the example of Maes (1994), as given in Clark (1997).

Chapter Ten: Babies and Bodies

1. Gallagher (2006).

2. Gallagher (2006).

3. Forssberg (1985).

4. Zelazo et al. (1972).

5. Zelazo (1983).

6. Thelen and Smith (1994); see also Thelen and Fisher (1982).

7. Thelen and Fisher (1982).

8. See, e.g., Super (1976).

9. Thelen and Smith (1994).

10. Clark (1997).

11. Clark (1997).

12. Smith (2005).

13. Adolph (1997, 1995); Adolph and Eppler (2002).

14. When toddlers walk, they propel themselves forward by falling downward while they stand on one foot, and then "catching" themselves with their moving foot (Adolph et al. [2003]). This is somewhat similar to Douglas Adams's description of learning to fly: "There is an art . . . or rather a knack to flying. The knack lies in learning how to throw yourself at the ground and miss."

15. Adolph (1997, 1995); Adolph and Eppler (2002).

16. Adolph (1997, 1995); Adolph and Eppler (2002).
17. Thelen et al. (2001).
18. Thelen et al. (2001); Smith (2005).
19. Smith (2005).
20. Smith (2005), p. 296.
21. Spelke and Hespos (2001); Durand and Lécuyer (2002); Hood et al. (2003).
22. Sheets-Johnstone (1999).
23. Smith and Gasser (2005); Smith (2005).
24. Smith (2005); Edelman (1987).
25. Meltzoff and Borton (1979).
26. Personally, I'm a bit skeptical of what looking-time experiments actually show: longer looking times have been argued to show both greater interest due to familiarity and also greater surprise due to novelty (!), but I suppose it is fair to say that they do indicate some form of discrimination on the part of infants, even if we can't fully characterize what or why that is.
27. Pfeifer and Bongard (2007).
28. Pfeifer and Bongard (2007).
29. Tucker and Ellis (1998).
30. Lakoff and Núñez (2001).
31. Pfeifer and Bongard (2007) call this the "principle of ecological balance."
32. See, for example, many of the papers in a special issue of Pragmatics & Cognition 17:3 (2009). It is also the case that words and their meanings are highly "culture-bound" by definition: Western language and modes of thought are very different from those of Eastern societies, for example. So this doesn't mean that all humans think the same, or that words never change their meaning over time. It simply means that language is a means to generate particular kinds of concepts that can be "frozen" in a format that gives them greater stability over time than the more embodied forms of representation that Smith considers.
33. Clark (1997).
34. This is very similar to Vygotsky's argument about public language as a psychological tool that could be used to bring one's own thoughts and actions under control.
35. See also Penn et al. (2008) for a similar take on this.
36. Clark (1997).

Chapter Eleven: Wider than the Sky

1. In case anyone is unfamiliar with the Segway, it looks like a scooter with the back cut off. It has two wheels and a long handlebar, and it is self-balancing so it doesn't tip over. Unlike a standard kid's scooter, it is motorized, so there's no pushing yourself along to speed up. Instead, the rider stands on a small platform between the wheels, moves by pushing the handlebar forward, and brakes by pulling it back. To turn left, one moves the handlebar to the left, and to turn right, one moves it to the right.

2. Clark (2008).
3. Wheeler (2005).
4. Clark and Chalmers (1998).
5. Lakoff and Johnson (1999).
6. For example, Adams and Aizawa (2010), p. 67, ask:

> Why did the pencil think 2 + 2 = 4?
> Clark's answer: Because it was coupled to the mathematician.

This assumes that, in order to count as part of a cognitive system, each part of the system must itself be cognitive and able to "think for itself." This is an obvious misunderstanding of the parity principle, which Clark illustrates with his own Q &A (Clark [2010], p. 82):

> Q. Why did the V4 neuron think that there was a spiral pattern in the stimulus?
> A. Because it was coupled to the monkey.

Part of the misunderstanding of Adams and Aizawa (2009, 2010) appears to derive from the fact that they fall foul of the mereological fallacy that we discussed in chapter 6. For example, they make statements like "The brain is capable of facial recognition" ([2009], p. 75), but, as we pointed out, brains don't recognize faces; only whole animals can. If one thinks of the brain as the seat of all cognitive processing, so giving the brain "a mind of its own," so to speak, then the idea of cognitive extension will indeed seem implausible. If, however, one sees the brain simply as one connection in the whole nervous-system-body-environment nexus that makes up a "mind," cognitive extension is obvious, because it is the system as whole that "thinks," and not the brain alone.

7. Clark (2008).
8. Clark (2008).
9. Maravita et al. (2003); Maravita and Iriki (2004).
10. Wilson and Clark (2009).
11. Bach-y-Rita et al. (1969); White et al. (1970); Bach-y-Rita and Kercel (2003); Bach-y-Rita et al. (2003).
12. Clark (2008).
13. Adams and Aizawa (2009).
14. Adams and Aizawa (2009). As we saw earlier, some of these criticisms of transcranialism seem to stem from the mereological fallacy.
15. Characterized by Wilson and Clark (2009) as the "dogma of intrinsic unsuitability" (p. 69).
16. Hurley (2010).
17. The tendency for people to recall the first and the last items in a list more accurately than those in the middle.
18. That is, if one is interested in understanding the specific brain mechanisms that underlie this functionalism, then knowledge of the constituent parts is useful and relevant to an understanding of the system as a whole.
19. Sutton (2010).

20. Pfeifer and Bongard (2007).

21. Ashby (1956).

22. Korriat and Goldsmith (1996).

23. Pfeifer and Bongard (2007).

24. This is largely due to the experimental strategy of "methodological solipsism" that is used in psychology and cognitive science, whereby the rest of the world beyond the individual is "bracketed off" during any attempt to characterize individual cognitive states and structures. It should be obvious that when one does this, the automatic assumption is being made that cognitive structures are things that occur solely in the head, and so it is not too surprising that we then find evidence that suggests exactly that . . .

25. Pfeifer and Bongard (2007).

26. Neisser (1982).

27. Neisser (1982).

28. Broca (1861), p. 237.

29. It might seem more appropriate to say "reconstructed," but Pfeifer and Bongard point out that this simply helps feed into the storehouse metaphor, because it suggests that there was a "thing" in the past that requires reconstructing in the present. Instead, there may never be any kind of "thing," only ongoing construction. This is obviously open to debate, but at least it gives us an alternative that is interesting to think about.

30. Ashby (1956).

31. Specifically, the birds were shown to be able to "plan for breakfast" (Raby et al. 2007). In the experiment, the birds first learned when they could expect to receive breakfast. They were fed powdered pine nuts (which can't be cached), went without food overnight, and then, in the morning, were placed in one of two rooms: in one of them, food was always provided; in the other, they were never given food. After three days of exposure to each room, the birds were given the opportunity to cache whole pine nuts. The experimenters reasoned that, if the birds were capable of "mental time travel"—that is, of projecting themselves into the future and understanding what that would be like—then they would cache food in the room in which they didn't receive breakfast, so that they could recover their cache and not go hungry. As predicted, the birds cached over three times as many pine nuts in the nonbreakfast room. One possibility, of course, is that the birds had learned to associate the nonbreakfast room with hunger, and so were more motivated to cache there. So, in a second experiment, the birds were fed every morning regardless of which room they were placed in, but the type of food differed: they were given dog kibble in one room, and pine nuts in the other. When the birds were given the chance to cache both pine nuts and dog kibble, they tended to cache more pine nuts in the dog kibble room, and more dog kibble in the pine nut room, so ensuring themselves a more varied breakfast.

These results were interpreted as a form of "future projection" and "mental time travel," but the experiments have been criticized for the way the results were analyzed (Suddendorf and Corballis [2008]), and because the second experiment lacked a control condition, so that it wasn't possible to distinguish whether the birds really were attempting to give themselves a more varied diet in the future as opposed to manifesting a simple preference for caching two different kinds of food, rather than only one (Premack [2007]).

32. Pfeifer and Bongard (2007).

33. Clark (1989).

34. Clark (2001); Triantafyllou and Triantafyllou (1995).

35. Clark (2001).

36. For more fascinating examples of how animals use the external world to aid in physiological processing, see Turner (2000).

37. Clark (1989, 2001).

38. Kirsh (1996).

39. Kirsh (1996).

40. Clark (2001).

41. Clark (1997, 2008).

42. Clark (1997).

43. Sterelny (2004b).

44. E.g., Ford (1991); Deecke et al. (2000); Yurk et al. 2002).

45. Di Fiore and Suarez (2007).

46. Strum et al. (1997).

REFERENCES

Adams, F., and Aizawa, K. (2009). Why the mind is still in the head. In: The Cambridge Handbook of Situated Cognition (eds. P. Robbins and M. Aydedem), pp. 75–96. Cambridge University Press, Cambridge.

Adams, F., and Aizawa, K. (2010). Defending the bounds of cognition. In: The Extended Mind (ed. R. Menary), pp. 67–80. MIT Press, Bradford Books, Cambridge, Mass.

Adolph, K. E. (1995). A psychological assessment of toddlers' ability to cope with slopes. Journal of Experimental Psychology: Human Perception and Performance 21: 734–750.

Adolph, K. E. (1997). Learning in the development of infant locomotion. Monographs of the Society for Research in Child Development 62 (Serial No. 251).

Adolph, K. E., and Eppler, M. (2002). Flexibility and specificity in infant motor skill acquisition. In: Progress in Infancy Research, vol. 2 (ed. J. Fagan), pp. 121–167. Erlbaum, Mahwah, N.J.

Adolph, K. E., Vereijken, B., and Scrout, P. E. (2003). What changes in infant walking and why? Child Development 74: 475–497.

Adolphs R. (2001). The neurobiology of social cognition. Current Opinion in Neurobiology 11: 231–239.

Alberts, J. R. (2007). Huddling by rat pups: ontogeny of individual and group behaviour. Developmental Psychobiology 49: 22–32.

Ashby, W. R. (1956). An Introduction to Cybernetics. Chapman and Hall, London.

Baader, A. P. (1996). The significance of visual landmarks for the navigation of the giant tropical ant, Paraponera clavata (Formicidae, Ponerinae). Insectes Sociaux 43: 435–450.

Bach-y-Rita, P., Collins, C. C., Saunders, F., White, B., and Scadden, L. (1969). Vision substitution by tactile image projection. Nature 221: 963–964.

Bach-y-Rita, P., and Kercel, S.W. (2003). Sensory substitution and the human-machine interface. Trends in Cognitive Science 7: 541–546.

Bach-y-Rita, P., Tyler, P. M., and Kaczmarek, K. (2003). Seeing with the brain. International Journal of Human-Computer Interaction 15: 285–295.

Ballerini, M., Cabibbo, N., Candelier, R., Cavagna, A., Cisbani, E., Giardina, I., Orlandi, A., Parisi, G., Procaccini, A., Viale, M., and Zdravkovic, V. (2008).

Empirical investigation of starling flocks: a benchmark study in collective animal behaviour. Animal Behaviour 76: 201–215.

Barrett, L. (2009). A guide to practical babooning: ecological, social and cognitive contingency. Evolutionary Anthropology. 18: 91–102.

Barrett, L. (in press). Why behaviourism isn't Satanism. In: The Oxford Handbook of Comparative Evolutionary Psychology (eds. J. Vonk and T. Shackelford). Oxford University Press, Oxford.

Barrett, L., and Henzi, S. P. (2005). The social nature of primate cognition. Proceedings of the Royal Society, London, Series B, 272: 1865–1875.

Barrett, L., Henzi, S. P., and Rendall, D. (2007). Social brains, simple minds: does social complexity really require cognitive complexity? Philosophical Transactions of the Royal Society, B, 362: 561–575.

Barton, R. A. (1996). Neocortex size and behavioural ecology in primates. Proceedings of the Royal Society, London, Series B, 263: 173–177.

Barton, R. A. (1998). Visual specialization and brain evolution in primates. Proceedings of the Royal Society, London, Series B, 265: 1933–1937.

Barton, R. A. (1999). The evolutionary ecology of the primate brain. In: Comparative Primate Socioecology (ed. P. C. Lee), pp.167–203. Cambridge University Press, Cambridge.

Barton, R. A. (2000). Ecological and social factors in primate brain evolution. In: On the Move: How and Why Animals Travel in Groups (eds. S. Boinski and P. Garber), pp. 204–237. University of Chicago Press, Chicago.

Barton, R.A. (2006). Primate brain evolution: integrating comparative, neurophysiological and ethological data. Evolutionary Anthropology 15: 224–236.

Bennett, M. R., and Hacker, P.M.S. (2003). Philosophical Foundations of Neuroscience. Blackwell Publishing, Oxford.

Blakemore, C. (1977) Mechanics of the Mind. Cambridge University Press, Cambridge.

Block, N. (1995). The mind as the software of the brain. In: An Invitation to Cognitive Science, 2nd edn. (eds. D. Osherson, L. Gleitman, S. Kosslyn, E. Smith, and S. Sternberg). MIT Press, Cambridge, Mass.

Blumberg, M. (2007). Anthropomorphism and evidence. Comparative Cognition and Behaviour Reviews 2: 145–146.

Blumberg, M. S., and Wasserman, E. A. (1995). Animal mind and the argument from design. American Psychologist 50: 133–144.

Bolhuis , J. J., and Honey, R. C. (1998). Imprinting, learning and development: from behaviour to brain and back. Trends in Neuroscience 31: 306–311.

Boroditsky, L. (2000). Metaphoric structuring: understanding time through spatial metaphors. Cognition 75: 1–28.

Bourbon, W. T. (1995). Perceptual control theory. In: Comparative Approaches to Cognitive Science (eds. H. L. Roitblat and J.-A. Meyer), pp. 151–171. MIT Press, Cambridge, Mass.

Broca, P. P. (1861). Loss of speech, chronic softening and partial destruction of the anterior left lobe of the brain. Bulletin de la Société d'Anthropologie 2: 235–238.

Brooks, R. A. (1999). Cambrian Intelligence: The Early History of the New A.I. MIT Press, Cambridge, Mass.

Brooks, R. A. (2002). Robot: The Future of Flesh and Machines. Allen Lane, London.

Brothers, L. (1990). The social brain: a project for integrating primate behaviour and neurophysiology in a new domain. Concepts in Neuroscience 1: 27–51.

Brothers, L., and Ring, B. (1992). A neuroethological framework for the representation of minds. Journal of Cognitive Neuroscience 4: 107–108.

Brothers, L., Ring, B., and Kling, A. (1992). Response of neurons in the macaque amygdala to complex social stimuli. Behavioral Brain Research 41: 199–213.

Buccino, G., Lui, F., Canessa, N., Patteri, I., Lagravenese, G., Benuzzi, F., Porro, C. A., and Rizzolatti, G. (2004). Neural circuits involved in the recognition of actions performed by non-conspecifics: an fMRI study. Journal of Cognitive Neuroscience 16: 114–126.

Byrne, R. W., and Bates, L. (2006). Why are animals cognitive? Current Biology 16: R445–R448.

Byrne, R. W., and Whiten, A. (1988). Machiavellian Intelligence: Social Expertise and the Evolution of Intellect in Monkeys, Apes and Humans. Clarendon Press, Oxford.

Cassia, V. M., Turati, C., and Simion, F. C. (2004). Can a non-specific bias toward top-heavy patterns explain newborn face preference? Current Directions in Psychological Science 15: 379–383.

Chemero, A. (2010). Radical Embodied Cognitive Science. MIT Press, Cambridge, Mass.

Churchland, P. (1986). Neurophilosophy. MIT Press, Cambridge, Mass.

Cisek, P. (1999). Beyond the computer metaphor: behaviour as interaction. In: Reclaiming Cognition: The Primacy of Action, Intention and Emotion (eds., R. Núñez and W. J. Freeman), pp. 125–142. Imprint Academic, Exeter.

Clark, A. (1989). Microcognition: Philosophy, Cognitive Science and Parallel Distributed Processing. MIT Press, Cambridge, Mass.

Clark, A. (1997). Being There: Putting Brain, Body, and World Together Again. MIT Press, Cambridge, Mass.

Clark, A. (2001). Mindware: An Introduction to the Philosophy of Cognitive Science. MIT Press, Cambridge, Mass.

Clark, A. (2008). Supersizing the Mind: Embodiment, Action and Cognitive Extension. MIT Press, Cambridge, Mass.

Clark, A. (2010). Coupling, constitution, and the cognitive kind: reply to Adams and Aizawa. In: The Extended Mind (ed. R. Menary), pp. 81–100. MIT Press, Bradford Books, Cambridge, Mass.

Clark, A., and Chalmers, D. (1998). The extended mind. Analysis 58: 7–19.

Clark, R. J., and Jackson, R. R. (1994). Self recognition in a jumping spider: Portia labiata females discriminate between their own draglines and those of conspecifics. Ethology Ecology and Evolution 6: 371–375.

Clark, R. J., and Jackson, R. R. (1995). Jumping spiders discriminate between the draglines of familiar and unfamiliar conspecifics. Ethology, Ecology and Evolution 7: 185–190.

Clark, R. J., Harland, D. P., and Jackson, R. R. (2000). Speculative hunting by an araneophagic salticid spider. Behaviour 137: 1601–1602.

Clayton, N. S., Dally, J. M., and Emery, N. J. (2007). Social cognition by food-caching corvids: the western scrub jay as a natural psychologist. Philosophical Transactions of the Royal Society,B, 362: 507–522.

Clayton, N. S., and Dickinson, A. (1998). Episodic-like memory during cache recovery by scrub jays. Nature 395: 272–274.

Cochin, S., Barthelemy, C., Roux, S., and Martineau, J. (1999). Observation and execution of movement: similarities demonstrated by quantified electroencephalograpy. European Journal of Neuroscience 11: 1839–1842.

Collins, S. H., Ruina, A., Tedrake, R., and Wisse, M. (2005). Efficient bipedal robots based on passive dynamic walkers. Science 307: 1082–1085.

Collins, S. H., Wisse, M., and Ruina, A. (2001). A three-dimensional passive dynamic walking robot with two legs and two knees. International Journal of Robotics Research 20: 607–615.

Costall, A. (1993). How Lloyd Morgan's canon backfired. Journal of the History of the Behavioral Sciences 29: 113–122.

Costall, A. (2004). From Darwin to Watson (and cognitivism) and back again: the principle of animal-environment mutuality. Behaviour and Philosophy 32: 179–195.

Costall, A., and Leudar, I. (2004). Where is the 'theory' in 'theory of mind'? Theory and Psychology 14: 623–646.

Costall, A., Leudar, I., and Reddy, V. (2006). Failing to see the irony in mind-reading. Theory and Psychology 16: 163.

Couzin, I. D., and Krause, J. (2003). Self-organization and collective behavior in vertebrates. Advances in the Study of Behavior 32: 1–75.

Crook, J. H. (1964). Field experiments on the nest construction and repair behaviour of certain weaver birds. Proceedings of the Zoological Society of London 142: 217–255.

Cruse, H. (1990). What mechanisms coordinate leg movement in walking arthropods? Trends in Neuroscience 13: 15–21.

Cziko, G. (2000). The Things We Do: Using the Lessons of Bernard and Darwin to Understand the What, How, and Why of Our Behavior. MIT Press, Cambridge, Mass.

Dally, J. M., Emery, N. J., and Clayton, N. S. (2004). Cache protection strategies by western scrub jays (Aphelocoma californica): hiding food in the shade. Proceedings of the Royal Society, Series B, 271: S837–S390.

Dally, J. M., Emery, N. J., and Clayton, N. S. (2005). Cache protection strategies by western scrub jays: implications for social cognition. Animal Behaviour 70: 1251–1263.

Dasser, V., et al. (1989). The perception of intention. Science 243: 365–367.

Dawkins, R. (1976). The Selfish Gene. Penguin, Harmondsworth.

Deecke, V. B., Ford, J.R.B., and Spong, P. (2000). Dialect change in resident killer whales: implications for vocal learning and cultural transmission. Animal Behaviour 40: 629–638.

Dennett, D. C. (1984). Elbow Room: The Varieties of Free Will Worth Wanting. MIT Press, Cambridge, Mass.

Dennett, D. C. (1989). The Intentional Stance. MIT Press, Cambridge, Mass.

Dennett, D. C. (2006). Breaking the Spell: Religion as Natural Phenomenon. Viking, New York.

de Waal, F.B.M. (1997). Are we in anthropodenial? Discover 18: 50–53.

de Waal, F.B.M. (2001). The Ape and the Sushi Master. Basic Books, New York.

de Waal, F.B.M. (2005). Animals and us: suspicious minds. New Scientist 186: 48.

Dewey, J. (1896). The reflex arc concept in psychology. Psychological Review 3: 357–370.

Di Fiore, A., and Suarez, S. A. (2007). Route-based travel and shared routes in sympatric spider and woolly monkeys: cognitive and evolutionary implications. Animal Cognition 10: 317–329.

Dittrich, W. H., and Lea, S. (1994). Visual perception of intentional motion. Perception 23: 253–268.

Dittrich, W. H., Troscianko, T., Lea, S.E.G., and Morgan, D. (1996). Perception of emotion from dynamic point-light displays represented in dance. Perception 25: 727–738.

Dorigo, M., Bonabeau, E., and Theraulaz, G. (2000). Ant algorithms and stigmergy. Future Generation Computer Systems 16: 851–871.

Draaisma, D. (2000). Metaphors of Memory: A History of Ideas about the Mind. Cambridge University Press, Cambridge.

Dreyfus, H.L. (1992). What Computers Still Can't Do. MIT Press, Cambridge, Mass.

Dreyfus, H. L.(2007). Why Heideggerian AI failed and how fixing it would require making it more Heideggerian. Artificial Intelligence 171: 1137–1160.

Dunbar, R.I.M. (1998). The social brain hypothesis. Evolutionary Anthropology 6: 178–190.

Durand, K., and Lécuyer, R. (2002). Object permanence observed in 4-month-old infants with a 2D display. Infant Behavior and Development 25: 269–278.

Dyer, A. G., Neumeyer, C., and Chittka, L. (2005). Honeybee (Apis mellifera) vision can discriminate between and recognize images of human faces. Journal of Experimental Biology

Dyer, J.R.G., Johansson, A., Helbing, D., Couzin, I. D., and Krause, J. (2009). Leadership, consensus decision making and collective behaviour in humans. Philosophical Transactions of the Royal Society, B, 364: 781–789.

Edelman, G. (1987). Neural Darwinism. Basic Books, New York.

Emery, N. J., and Clayton, N. S. (2001). Effects of experience and social context on prospective caching strategies in scrub jays. Nature 414: 443–446.

Fadiga, L., Fogassi, L., Pavesi, G., and Rizzolatti, G. (1995). Motor facilitation during action observation: a magnetic stimulation study. Journal of Neurophysiology 73: 2608–2611.

Fodor, J. (1999). Not so Clever Hans? London Review of Books, February 4.

Ford, J.K.B. (1991). Vocal traditions among resident killer whales (Orcinus orca) in coastal waters of British Columbia. Canadian Journal of Zoology 69: 1454–1483.

Forssberg, H. (1985). Ontogeny of human locomotor control. I. Infant stepping, supported locomotion, and transition to independent locomotion. Experimental Brain Research 57: 480–493.

Freeman, W. J. (1991). The physiology of perception. Scientific American 264: 78–85.

Freeman, W. J. (1995). Societies of Brains: A Study in the Neuroscience of Love and Hate. Erlbaum, Mahwah, N.J.

Freeman, W. J. (2000). How Brains Make Up Their Minds. Phoenix, London.

Freeman, W. J., and Skarda, C. A. (1990). Mind/brain science: neuroscience on philosophy of mind. In: John Searle and His Critics. (eds. E. Lepore and R. van Gulick), pp. 115–127. Blackwell, Oxford.

Gallagher, S. (2006). How the Body Shapes the Mind. Oxford University Press, Oxford.

Gallese, V. (2001). The 'shared manifold' hypothesis: from mirror neurons to empathy. Journal of Consciousness Studies 8: 33–50.

Gallese, V. (2005). Embodied simulation: from neurons to phenomenal experience. Phenomenology and Cognitive Science 4: 22–48.

Gallese, V., Fadiga, L., Fogassi, L., and Rizzolatti, G. (1996). Action recognition in the premotor cortex. Brain 119: 593–609.

Gauthier, I., Skudlaski, P., Gore, J. C., and Anderson, A. W. (2000). Expertise for cars and birds recruits brain areas involved in face recognition. Nature Neuroscience 3: 191–197.

Gergely, G., et al. (1995). Taking the intentional stance at 12 months of age. Cognition 56: 165–193.

Gibson, J. J. (1950). The Perception of the Visual World. Houghton-Mifflin, Boston.

Gibson, J. J. (1966). The Senses Considered as Perceptual Systems. Houghton Mifflin, Boston.

Gibson, J. J. (1970). On the relation between hallucination and perception. Leonardo 3: 425–427.

Gibson, J. J. (1974). Visualizing conceived as visual apprehending without any particular point of observation. Leonardo 7: 41–42.

Gibson, J. J. (1979). The Ecological Approach to Visual Perception. Erlbaum, Mahwah, N.J..

Gilbert, S. F. (2005). Mechanisms for the regulation of gene expression: ecological aspects of animal development. Journal of Bioscience 30: 101–110.

Goren, C. C., Sarty, M., and Wu, P.Y.K. (1975). Visual following and pattern discrimination of face-like stimuli by newborn infants. Pediatrics 56: 544–549.

Gottleib, G. (1971). Development of Species Identification in Birds. University of Chicago Press, Chicago.

Gottleib, G. (1981). Roles of early experience in species-specific perceptual development. In: Development of Perception, vol. 1 (eds. R. N. Astline, J. R. Alberts, and M. R. Petersen), pp. 5–54. Academic Press, San Diego.

Gottlieb, G. (1985). Development of species identification in ducklings XI: embryonic critical period for species-typical perception in the hatchling. Animal Behaviour 33: 225–233.

Gottlieb, G. (1991). Experiential canalization of behavioral development: results. Developmental Psychology 27: 35–39.

Granzow, J. E., (2010). Ventriloquial Dummy Tones: Embodied Cognition of Pitch Perception. Unpublished MSc. Thesis, University of Lethbridge.

Gregory, R. L. (1980). Hypotheses as perceptions. Philosophical Transactions of the Royal Society, London, Series B, 290: 181–197.

Grey Walter, W. (1950). An imitation of life. Scientific American 182: 42–45.

Grey Walter, W. (1951). A machine that learns. Scientific American 185: 60–63.

Grey Walter, W. (1953). The Living Brain. W. W. Norton, New York.

Grèzes J., Armony, J. L., Rowe, J., and Passingham. R. E. (2003). Activations related to "mirror" and "canonical" neurones in the human brain: an fMRI study. Neuroimage 18: 928–937.

Grill-Spector, K., Knouf, N., and Kanwisher, N. (2004). The fusiform face area subserves face perception, not generic within-category identification. Nature Neuroscience 7: 555–562.

Gubernick, D. J., and Alberts, J. R. (1983). Maternal licking of young: Resource exchange and proximate controls. Physiology & Behavior 31: 593–601.

Guthrie, S. (1993). Faces in the Clouds: A New Theory of Religion. Oxford University Press, Oxford.

Hacker, P.M.S. (1991). Experimental methods and conceptual confusion: an investigation into R. L. Gregory's theory of perception. Iyyun: The Jerusalem Philosophical Quarterly 40: 289–314.

Hacker, P.M.S. (1995). Helmholtz's theory of perception: an investigation into its conceptual framework. International Studies in the Philosophy of Science 9: 199–214.

Hari, R., Forss, N., Avikainen, S., Kirveskari, S., Salenius, S., and Rizzolatti, G. (1998). Activation of human primary motor cortex during action observation: a neuromagnetic study. Proceedings of the National Academy of Science USA 95: 15061–15065.

Harland, D. P., and Jackson, R. R. (2000). "Eight-legged cats" and how they see—a review of recent research on jumping spiders (Aranae: Salticidae). Cimbebasia 16: 231–240.

Harland, D. P., and Jackson, R. R. (2004). Portia perceptions: the Umwelt of an araneophagic jumping spider. In: Complex Worlds from Simpler Nervous Systems (ed. F. Prete), pp. 5–40. MIT Press, Bradford Books, Cambridge, Mass.

Hashimoto, H. (1966). A phenomenal analysis of social perception. Journal of Child Development 2: 3–26.

Haugeland, J. (1995). Mind embodied and embedded. In: Having Thought: Essays in the Metaphysics of Mind, pp. 207–240. Harvard University Press, Cambridge, Mass.

Hauser, M. (1988). A non-human primate's expectations about object motion and destination: the importance of self-propelled movement and animacy. Developmental Science 1: 31–38.

Haxby, J. V., Gobbini, M. I., Furey, M. L., Ishai, A., Schouten, J. L., and Pietrini, P. (2001). Distributed and overlapping representations of faces and objects in ventral temporal cortex. Trends in Cognitive Science 10: 278–285.

Hayward, R. (2001a). The tortoise and the love machine: Grey Walter and the politics of electroencephalograhy. In: Mindful Practices: On the Neurosciences in the Twentieth Century 14, no. 4. (eds. C. Borck and E. Hanger), pp. 615–641.

Hayward, R. (2001b). Our friends electric: mechanical models of the mind in post-war Britain. In: A Century of Psychology (eds. G. Bunn, A. Lovie, and G. Richards), pp. 290–308. British Psychological Society.

Hebb, D. (1949). The Organization of Behavior. Wiley, New York.

Hedwig, B., and Webb, B. (2005). Mechanisms underlying phonotactic steering in the cricket, Gryllus bimaculatus, revealed with a fast trackball system. Journal of Experimental Biology 208: 915–927.

Heidegger, M. (1927). Being and Time. Harper, New York.

Heider, F., and Simmel, M. (1944), An experimental study of apparent motion. American Journal of Psychology 57: 243–249.

Heil, K. H. (1936). Beitrage zur physiologie und psychologie der Springspinnen. Zeitschrift für Vergleichende Physiologie 23: 125–149.

Helmholtz, H. (1868). The recent progress of the theory of vision. Repr. in translation in: Helmholtz on Perception (eds. R. M. Warren and R. P. Warren), p. 101. Wiley, New York.

Henzi, S. P., and Barrett, L. (2003). Evolutionary ecology, sexual conflict and behavioral differentiation among baboon populations. Evolutionary Anthropology 12: 217–230.

Henzi, S. P., and Barrett, L. (2005). The historical socio-ecology of savannah baboons. Journal of Zoology, London, 265: 215–226.

Hess, E. H. (1958). Imprinting in animals. Scientific American 198: 81–90.

Hesse, M. B. (1966). Models and Analogies in Science. University of Notre Dame Press, Notre Dame, Ind.

Heyes, C. M. (1998). Theory of mind in nonhuman primates. Behavioral and Brain Sciences 21: 101–148.

Hill, D. E. (1979). Orientation by jumping spiders of the genus Phidippus pulcherrimus (Araneae: Salticidae) during the pursuit of prey. Behavioral Ecology and Sociobiology 5: 301–322.

Holland, O. (2003). The first biologically inspired robots. Robotica 21: 351–363.

Holyoak, K., and Thagard, P. (1996). Mental Leaps: Analogy in Creative Thought. MIT Press, Cambridge, Mass.

Hood, B., Cole-Davies, V., and Dias, M. (2003). Looking and search measures of object knowledge in preschool children. Developmental Psychology 39: 61–70.

Horchler, A. D., Reeve, R. R., Webb, B., and Quinn, R. D. (2004). Robot phonotaxis in the wild: a biologically inspired approach to outdoor sound localisation. Advanced Robotics 18: 801–816.

Howard, M. (2009). Three years in jail for journalist who threw shoe at Bush. http://www.guardian.co.uk/world/2009/mar/13/journalist-shoe-bush-jail".

Hughes, J. F., Skaletsky, H., Pyntikova, T., Graves, T. A., Van Daalen, S.K.M., Minx, P. J., Fulton, R. S., McGrath, S. D., Locke, D. P., Friedman, C., Trask, B. J., Mardis, E. R., Warren, W. C., Repping, S., Rozen, S., Wilson, R. K., and Page, D. C. (2010). Chimpanzee and human Y chromosomes are remarkably divergent in structure and gene content. Nature 463: 536–539.

Hume, D. (1889). The Natural History of Religion. Bradlaugh Bonner, London.

Hurley, S. (2010). The varieties of externalism. In: The Extended Mind (ed. R. Menary), pp. 101–154. MIT Press, Bradford Books, Cambridge, Mass.

Hyman, J. (1989). The Imitation of Nature. Basil Blackwell, Oxford.

Iida, F., and Pfeifer, R. (2004). Self-stabilization and behavioural diversity of embodied adaptive locomotion. In: Embodied Artificial Intelligence, LNAI 3139 (eds. F. Iida, R. Pfeifer, L. Steels, and Y. Kuniyoshi), pp. 119–129. Springer-Verlag, Berlin.

Inoue, S., and Matsuzawa, T. (2007). Working memory of numerals in chimpanzees. Current Biology 17: R1004–R1005.

Jackson, R. R., and Wilcox, R.S. (1990). Aggressive mimicry, prey-specific predatory behaviour and predator recognition in the predator-prey interactions of Portia fimbriata and Euryattus sp. jumping spiders from Queensland. Behavioral Ecology and Sociobiology 26: 111–119.

Jackson, R. R., and Wilcox, R. S. (1994). Spider flexibly chooses aggressive mimicry signals for different prey by trial and error. Behaviour 127: 21–36.

Johansson, G. (1973). Visual perception of biological motion and a model for its analysis. Perception and Psychophysics 14: 201– 211.

Johnson, M. H., Dziurawiec, S., Ellis, H. D.. and Morton, J. (1991). Newborns' preferential tracking of faces and its subsequent decline. Cognition 40: 1–19.

Johnson, M., and Rohrer, T. (2007). We are live creatures: embodiment, American Pragmatism, and the cognitive organism. In: Body, Language, and Mind, vol. 1 (eds. J. Zlatev, T. Ziemke, R. Frank, and R. Dirven), pp. 17–54. Mouton de Gruyter, Berlin.

Kandel, E. R., Schwartz, J. H., and Jessell, T. M. (1995). Essentials of Neural Science and Behavior. Appleton and Lang, Norwalk, Conn.

Kanwisher, N., McDermott, J., and Chun, M. (1997). The fusiform face area: a module in human extrastriate cortex specialized for the perception of faces. Journal of Neuroscience 17: 4302–4311.

Kelso, J.A.S. (1995). Dynamic Patterns: The Self-Organization of Brain and Behavior. Bradford Books, MIT Press, Cambridge, Mass.

Kennedy, J. S. (1992). The New Anthropomorphism. Cambridge University Press, Cambridge.

Killeen, P. R., and Glenberg, A. M. (2010). Resituating cognition. Comparative and Biobehavioral Reviews 5: 59–77.

Kirsh, D. (1996). Adapting the environment instead of oneself. Adaptive Behavior 4: 415–452.

Koriat, A., and Goldsmith, M. (1996). Memory metaphors and the real-life/laboratory controversy: correspondence versus storehouse conceptions of memory. Behavioral and Brain Sciences 19: 167–228.

Kozlowski, L. T., and Cutting, J. E. (1977). Recognizing the sex of a walker from a dynamic point light display. Perception and Psychophysics 21: 575–580.

Lakoff, G., and Johnson, M. (1999). Philosophy in the Flesh: The Embodied Mind and the Challenge to Western Thought. Harper-Collins, London.

Lakoff, G., and Núñez, R. (2001). Where Mathematics Comes From: How the Embodied Mind Brings Mathematics into Being. Basic Books, New York.

Land, M. F. (1969a). Structure of the retinae of the eyes of jumping spiders (Salticidae: Dendryphantinae) in relation to visual optics. Journal of Experimental Biology 51: 443–470.

Land, M. F. (1969b). Movements of the retinae of jumping spiders (Salticidae: Dendryphantinae) in response to visual stimuli. Journal of Experimental Biology 51: 471–493.

Land, M. F. (1972). Stepping movements made by jumping spiders during turns mediated by lateral eyes. Journal of Experimental Biology 57: 15–40.

Land, M. F., and Nilsson, D.-E. (2002). Animal Eyes. Oxford University Press, New York.

Landauer, T. K., and Dumais, S. T. (1997). A solution to Plato's problem: the latent semantic analysis theory of acquisition, induction and representation of knowledge. Psychological Review 104: 211–240.

Leary, D. E. (ed.). (1990). Metaphors in the History of Psychology. Cambridge University Press, Cambridge.

Lefebvre, L., Whittle, P., Lascaris, E., and Finkelstein, A. (1997). Feeding innovations and forebrain size in birds. Animal Behaviour 53: 549–560.

Le Grand, R., Mondloch, C. J., Maurer, D., and Brent, H. P. (2001a). Early visual experience and face processing. Nature 410: 890.

Le Grand, R., Mondloch, C. J., Maurer, D., and Brent, H. P. (2001b). Correction: Early visual experience and face processing. Nature 412: 786.

Le Grand, R., Mondloch, C. J., Maurer, D., and Brent, R. P. (2004). Impairment in holistic face processing following early visual deprivation. Psychological Science 15: 762–768.

Leudar, I., and Costall, A. (2004). On the persistence of the 'problem of other minds' in psychology: Chomsky, Grice and 'Theory of Mind.' Theory and Psychology 14: 601–621.

Lichtensteiger, L., and Salomon, R. (2000). The evolution of an artificial compound eye by using adaptive hardware. In: Proceedings of the 2000 Congress on Evolutionary Computation, pp. 1144–1151. IEEE Press, Piscataway, N.J.

Llinás, R. (1987). "Mindfulness" as a functional state of the brain. In: Mindwaves (eds. C. Blakemore and S. Greenfield), pp. 339–358. Basil Blackwell, Oxford.

Lorenz, K. (1937). The companion in the bird's world. Auk 54: 245–273.

Lund, H. H., Webb, B., and Hallam, J. (1997). A robot attracted to the cricket species Gryllus bimaculatus. In: Proceedings of the Fourth European Conference on Artificial Life (eds. P. Husbands and I. Harvey), pp. 246–255. MIT Press, Cambridge, Mass.

Lund, H. H., Webb, B., and Hallam, J. (1998). Physical and temporal scaling considerations in a robot model of cricket calling song preference. Artificial Life 4: 95–107.

Macfarlane A. (1975). Olfaction in the development of social preferences in the human neonate. In: Parent-Infant Interactions (eds. R. Porter R and M. O'Connor), pp. 103–113. Ciba Foundation Symposium 33. Elsevier, New York.

Macrae, F. (2009). Stone the crows! Santino the rock-throwing ape proves chimps plan head. Daily Mail, March 12. http://www.dailymail.co.uk/sciencetech/article-1160681/Stone-crows-Santino-rock-throwing-ape-proves-chimps-plan-ahead.html#ixzz0bUblgYS.

Maes, P. (1994). Modeling adaptive autonomous agents. Artificial Life 1: 135–162.

Makin, J. W., and Porter, R. H. (1989). Attractiveness of lactating females' breast odors to neonates. Child Development 60: 803–810.

Malone, J. (2009). Psychology: Pythagoras to Present. MIT Press, Cambridge, Mass.

Mameli, M. (2001). Mind reading, mind shaping and evolution. Biology and Philosophy 16: 595–626.

Maravita, A., and Iriki, A. (2004). Tools for the body (schema). Trends in Cognitive Science 8: 79–86.

Maravita, A., Spence, C., and Driver, J. (2003). Multisensory integration of the body schema: close at hand and within reach. Current Biology 13: R531–R539.

Maris, M., and te Boekhorst, R. 1996. Exploiting physical constraints: heap formation through behavioral error in a group of robots. In: Proceedings of IROS '96: IEEE/RSJ International Conference on Intelligent Robots and Systems (ed. M. Asada), pp. 1655–1660. IEEE Press, Piscataway, N.J.

Marr, D. (1982). Vision: A Computational Investigation into the Human Representation and Processing of Visual Information. W. H. Freeman, San Francisco.

Mataric, M. J. (1990). Navigating with a rat brain: a neurobiologically-inspired model for robot spatial representation. In: Proceedings of the First International

Conference on Simulation of Adaptive Behavior (eds. J. Meyer and S. Wilson), pp. 169–175. MIT Press, Bradford Books, Cambridge, Mass.

Mather, G., and West, S. (1993). Recognition of animal locomotion from dynamic point-light displays. Perception 22: 759–766.

Maturana, H. R., and Varela, F. J. (1998). The Tree of Knowledge: the Biological Roots of Human Understanding. Shambhala, London.

Maurer, D., Le Grand, R., and Mondloch, C. J. (2002). The many faces of configural processing. Trends in Cognitive Sciences 6: 255–260.

May, C. J., Schank, J. C., Joshi, S., Tran, J., Taylor, R. J. and Scott, I.(2006). Rat pups and random robots generate similar self-organized and intentional behavior. Complexity 12: 53–66.

McComb, K., Moss, C., Durant, S. M., Baker, L., and Sayialel, S. (2001). Matriarchs as repositories of social knowledge in African elephants. Science 292: 491–494.

McGeer, T. (1990). Passive dynamic walking. International Journal of Robotics Research 9: 62–82.

Meltzoff, A., and Borton, R. W. (1979). Intermodal matching by human neonates. Nature 282: 403–404.

Michaels, C. F., and Carello, C. (1981). Direct Perception. Prentice Hall, Englewood Cliffs, N.J.

Michelsen, A., Popov, A. V., and Lewis, B. (1994). Physics of directional hearing in the cricket, Gryllus bimaculatus. Journal of Comparative Physiology A 175: 153–164.

Miller, G. A. (2003). The cognitive revolution: a historical perspective. Trends in Cognitive Sciences 7: 141–144.

Morgan, C. L. (1890). Animal Life and Intelligence. London: Edward Arnold.

Morgan, C. L. (1894). An Introduction to Comparative Psychology. Walter Scott, Ltd., London.

Morris, M. W. and Peng, K. (1994). Culture and cause: American and Chinese attributions for social and physical events. Journal of Personality and Social Psychology 67: 949–971.

Mossio, M., and Taraborelli, D. (2008). Action-dependent perceptual invariants: from ecological to sensorimotor approaches. Consciousness and Cognition 17: 1324–1340.

Neisser, U. (1982). Memory: what are the important questions? In: Practical Aspects of Memory (eds. M. M. Gruneberg, P. E., Maris, and R. N. Sykes), pp. 3–24. Academic Press, San Diego, Calif.

Newell, A., and Simon, H. (1974). Computer science as empirical inquiry: symbols and search. Communications of the ACM 19: 113–126.

Noë, A. (2004). Action in Perception. MIT Press, Cambridge, Mass.

Noë, A. (2009). Out of Our Heads: Why You Are Not Your Brain, and Other Lessons from the Biology of Consciousness. Hill and Wang, New York.

Odling-Smee, F. J., Laland, K. N., and Feldman, M. W. (2003). Niche Construction: The Neglected Process in Evolution. Princeton University Press, Princeton, N.J.

Ortony, A. (ed.). (1979) Metaphor and Thought. Cambridge University Press, Cambridge.

Osvath, M. (2009). Spontaneous planning for future stone-throwing by a male chimpanzee. Current Biology 19: R190–R191.

Overington, S. E., Morand-Ferron, J., Boogert, N. J., and Lefebvre, L. (2009). Technical innovations drive the relationship between innovativeness and residual brain size in birds. Animal Behaviour 78: 1001–1010.

Oyama, S. (1985). The Ontogeny of Information. MIT Press, Cambridge, Mass.

Oyama, S. (1989). Ontogeny and the central dogma: do we need the concept of genetic programming in order to have an evolutionary perspective? In: Systems and Development: The Minnesota Symposia on Child Psychology, vol. 22, (eds. M. R. Gunnar and E. Thelen), pp. 1–34. Erlbaum, Hillsdale, N.J.

Panksepp, J. (1998). Affective Neuroscience: The Foundations of Human and Animal Emotions. Oxford University Press, New York.

Penn, D. C., Holyoak, K. J., and Povinelli, D. J. (2008). Darwin's mistake: explaining the discontinuity between human and nonhuman minds. Behavioral and Brain Sciences 31: 109–178.

Perrett, D. I. (1999). A cellular basis for reading minds from faces and actions. In Behavioral and Neural Mechanisms of Communication (eds. M. Hauser and M. Konishi), pp. 159–185. MIT Press, Cambridge, Mass.

Perrett, D. I., Hietanen, J. K., Oram, M. W., and Benson, P. J. (1992). Organization and functions of cells responsive to faces in the temporal cortex. Philosophical Transactions of the Royal Society of London, B, 335: 23–30.

Perrett, D. I., Mistlin, A. J., and Chitty, A. J. (1987). Visual cells responsive to faces. Trends in Neurosciences 10: 358–364.

Perrett, D. I., Rolls, E. T., and Caan, W. (1982). Visual neurones responsive to faces in the monkey temporal cortex. Experimental Brain Research 47: 329–342.

Perrett, D. I., Smith, P.A.J., Potter, D. D., et al. (1984). Neurones responsive to faces in the temporal cortex: studies of functional organization, sensitivity to identity and relation to perception. Human Neurobiology 3: 197–208.

Perrett, D. I., Smith, P.A.J., Potter, D. D., et al. (1985). Visual cells in the temporal cortex sensitive to face view and gaze direction. Proceedings of the Royal Society of London, Series B, 223: 293–317.

Pfeifer, R., and Bongard, J. (2007). How the Body Shapes the Way We Think. MIT Press, Cambridge, Mass.

Pfeifer, R., and Scheier, C. 1999. Understanding Intelligence. MIT Press, Cambridge, Mass.

Pickering, A. (2010). The Cybernetic Brain: Sketches of Another Future. University of Chicago Press, Chicago.

Pinker, S. (2003). How the Mind Works. Penguin, London.

Plotkin, H. (1994). Darwin Machines and the Nature of Knowledge. Penguin, Harmondsworth.

Plotkin, H. (1998). Evolution in Mind. Penguin, Harmondsworth.

Porter, R. H., Makin, J. W., Davis, L. B., and Christensen, K. M. (1992). Breast-fed infants respond to olfactory cues from their own mother and unfamiliar lactating females. Infant Behavior and Development 15: 85–93.

Porter, R. H., and Winberg, J. (1999). Unique salience of maternal breast odors for newborn infants. Neuroscience and Biobehavioral Reviews 23: 439–449.

Powers, W. T. (1973). Behavior: The Control of Perception. Aldine, Chicago.

Premack, D. (2007). Human and animal cognition: continuity and discontinuity. Proceedings of the National Academy of Sciences 104: 13861–13867.

Quinn, R. D, Nelson, G. M., Bachman, R. J., Kingsley, D. A., Offi, J., and Ritzmann, R. E. (2001). Insect designs for improved robot mobility. In: Proceedings of Fourth International Conference on Climbing and Walking Robots (eds. K. Berns and R. Dillman), pp. 69–76. Professional Engineering Publishing, London.

Raby, C. R., Alexis, D. M., Dickinson, A., and Clayton, N. S. (2007). Planning for the future by western scrub-jays. Nature 445: 919–921.

Reader, S. M., and Laland, K. N. (2002). Social intelligence, innovation, and enhanced brain size in primates. Proceedings of the National Academy of Science USA 99:4436–4441.

Reddy, V., and Morris, P. (2004). Participants don't need theories: knowing minds in engagement. Theory and Psychology 14: 647–665.

Reed, E. S. (1996). Encountering the World: Toward an Ecological Psychology. Oxford University Press, Oxford.

Rizzolatti, G., Fadiga, L., Gallese, V. and Fogassi, L. (1996). Premotor cortex and recognition of motor actions. Cognitive Brain Research 3: 131–141.

Rockwell, W. T. (2005). Neither Brain nor Ghost: A Nondualist Alternative to the Mind-Brain Identity Theory. MIT Press, Cambridge, Mass.

Rowlands, M. (2003). Externalism: Putting Mind and World Back Together Again. McGill-Queen's University Press, Montreal and Kingston.

Sample, I. (2009). Chimp who threw stones at zoo visitors showed human trait, says scientist. Guardian, March 9. http://guardian.co.uk/science/2009/mar/09/chimp-zoo-stones-science.

Schaal, B., Marlier, L., and Soussignan, R. (2000). Humans learn odours from their pregnant mother's diet. Chemical Senses 25: 729–737.

Schank, J. C., and Alberts, J. R. (1997). Self-organized huddles of rat pups modeled by simple rules of individual behaviour. Journal of Theoretical Biology 189: 11–25.

Schank, J. C., and Alberts, J. R. (2000). The developmental emergence of coupled activity as cooperative aggregation in rat pups. Proceedings of the Royal Society, London, Series B, 267: 2307–2315.

Scholl, B. J., and Tremoulet, P. D. (2000). Perceptual causality and animacy. Trends in Cognitive Science 8: 299–309.

Schuck-Paim, C., Alonso, W. J., and Ottoni, E. B. (2008). Cognition in an ever-changing world: climatic variability is associated with brain size in neotropical parrots. Brain, Behaviour and Evolution 71: 200–215.

Searle, J. R. (1983). Intentionality: An Essay in the Philosophy of Mind. Cambridge University Press, Cambridge.

Seither-Preisler, A., Johnson, L., Krumbholz, K., Nobbe, A., Patterson, R., and Seither, S. (2007). Tone sequences with conflicting fundamental pitch and timbre changes are heard differently by musicians and nonmusicians. Journal of Experimental Psychology: Human Perception and Performance 33: 743–751.

Sheets-Johnstone, M. (1999). Emotion and movement: a beginning empirical-phenomenological analysis of their relationship. Journal of Conciousness Studies 6: 259–271.

Shepard, R. N. (1984). Ecological constraints on internal representation: resonant kinematics of perceiving, imagining, thinking and dreaming. Psychological Review 91: 417–447.

Silberberg, A., and Kearns, D. (2009). Memory for the order of briefly presented numerals in humans as a function of practice. Animal Cognition 12: 405–407.

Simion, F., Valenza, E., Cassia, V. M., Turati, C., and Umilta, C. (2002). Newborns' preference for up-down asymmetrical configurations. Developmental Science 5: 427–434.

Simons, H. A. (1996). The Sciences of the Artificial, 3rd edn.. MIT Press, Cambridge, Mass.

Skarda, C. A., and Freeman, W. J. (1990). Chaos and the new science of the brain. Concepts in Neuroscience 1: 275–285.

Skinner, B. F. (1987). Whatever happened to psychology as the science of behavior? American Psychologist 42: 780–786.

Smith, L. B. (2005). Cognition as a dynamic system: principles from embodiment. Developmental Review 25: 278–298.

Smith, L. B., and Gasser, M. (2005). The development of embodied cognition: six lessons from babies. Artificial Life 11: 13–30.

Sol, D., Bacher, S., Reader, S. M., and Lefebvre, L. (2008). Brain size predicts the success of mammal species introduced into novel environments. American Naturalist 172: 563–571.

Sol, D., Duncan, R. P., Blackburn, T. M., Cassey, P., and Lefebvre, L. (2005). Big brains, enhanced cognition and response of birds to novel environments. Proceedings of the National Academy of Sciences USA 102: 5460–5465.

Spelke, E., and Hespos, S. (2001). Continuity, competence, and the object concept. In: Language, Brain, and Cognitive Development: Essays in Honor of Jacques Mehler (ed. E. Dupoux), pp. 325—340. MIT Press, Cambridge, Mass.

Sperry, R. W. (1945). Restoration of vision after crossing of optic nerves and after contralateral transplantation of eye. Neurophysiology 8: 15–28.

Sperry, R. W. (1951). Mechanisms of neural maturation. In: Handbook of Experimental Psychology (ed. S. S. Stevens), pp. 236–280. John Wiley, New York.

Spivey, M. (2007). The Continuity of Mind. Oxford University Press, Oxford.

Sterelny, K. (2004a). Symbiosis, evolvability and modularity. In: Modularity in Development and Evolution (eds. G. Schlosser and G. Wagner), pp. 490–516. University of Chicago Press, Chicago.

Sterelny, K. (2004b). Externalism, epistemic artefacts and the extended mind. In: The Externalist Challenge (ed. R. Schantz), pp. 239–254. Walter de Gruyter, Berlin.

Strum, S. C., Forster, D., and Hutchins, E. (1997). Why Machiavellian intelligence may not be Machiavellian. In Machiavellian Intelligence II: Extensions and Evaluations (eds. A. Whiten and R. W. Byrne), pp. 50–85. Cambridge University Press, Cambridge.

Suddendorf, T., and Corballis, M. C. (2008). New evidence for animal foresight? Animal Behaviour 75: e1–e3.

Super, C. (1976). Environmental effects on motor development: the case of "African infant precocity." Developmental Medicine and Child Neurology 18: 561–567.

Sutton, J. (2010). Exograms and interdisciplinarity: history, the extended mind and the civilizing process. In: The Extended Mind (ed. R. Menary), pp. 189–226. MIT Press, Bradford Books, Cambridge, Mass.

Tanaka, J., and Farah, M. (1993). Parts and wholes in face recognition. Quarterly Journal of Experimental Psychology 46: 225–245.

Tarr, M. J., and Gauthier, I. (2000). FFA: a flexible fusiform area for subordinate level processing automated by expertise. Nature Neuroscience 3: 764–769.

Tarsitano, M. (2006). Route selection by a jumping spider (Portia labiata) during the locomotory phase of a detour. Animal Behaviour 72: 1437–1442.

Tarsitano, M. S., and Andrew, R. (1999). Scanning and route selection in the jumping spider, Portia labiata. Animal Behaviour 58: 255–265.

Tarsitano, M. S., and Jackson, R. R. (1997). Araneophagic jumping spiders discriminate between detour routes that do and do not lead to prey. Animal Behaviour 53: 257–266.

Tebbich, S., Taborsky, M., Fessl, B., and Blomqvist, D. (2001). Do woodpecker finches acquire tool-use by social learning? Proceedings of the Royal Society, Series B, 268: 2189–2193.

Thach, W., Goodkin, H., and Keating, J. (1992). The cerebellum and the adaptive coordination of movement. Annual Review of Neuroscience 15: 403–442.

Thelen, E., and Fisher, D. M. (1982) Newborn stepping: an explanation for a 'diappearing' reflex. Developmental Psychology 18: 760–775.

Thelen, E., Schöner, G., Scheier, C., and Smith, L. B. (2001). The dynamics of embodiment: a field theory of infant perseverative reaching. Behavioral and Brain Sciences 24: 1–86.

Thelen, E., and Smith, L. (1994). A Dynamic Systems Approach to the Development of Cognition and Action. MIT Press, Cambridge, Mass.

Tibbets, E. A. (2002). Visual signals of individual identity in the wasp Polistes fuscata. Proceedings of the Royal Society of London, Series B, 269: 1423–1428.

Tinbergen, N. (1969). The Study of Instinct. Oxford University Press, Oxford.

Tomasello, M. (1999). The Cultural Origins of Human Cognition. Harvard University Press, Cambridge, Mass.

Tooby, J., and Cosmides, L. (2005). Conceptual foundations of evolutionary psychology. In: The Handbook of Evolutionary Psychology (ed. D. Buss), pp. 5–67. Wiley, New York.

Tremoulet, P. D., and Feldman, J. (2000). Perception of animacy from motion of a single object. Perception 29: 943–951.

Triantafyllou, M., and Triantafyllou, G. (1995). An efficient swimming machine. Scientific American 272: 64–70.

Tucker, M., and Ellis, R. (1998). On the relation between seen objects and components of potential actions. Journal of Experimental Psychology: Human Perception and Performance 24: 830–846.

Turner, J. S. (2000). The Extended Organism. Harvard University Press, Cambridge, Mass.

Tyler, T. (2003). If horses had hands . . . Society and Animals 11: 267–281.

Uller. C. (2004). Disposition to recognize goals in infant chimpanzees. Animal Cognition 7: 154–161.

Van Gelder, T. (1995). What might cognition be, if not computation? Journal of Philosophy 92: 345–381.

Varela, F. J., Thompson, E., and Rosch, E. (1991). The Embodied Mind: Cognitive Science and Human Experience. MIT Press, Cambridge, Mass.

Varendi, H., Porter, R.H., and Winberg, J. (1996). Attractiveness of amniotic fluid odor: evidence of prenatal olfactory learning? Acta Paediatrica 85: 1223–1227.

Varendi, H., Porter, R. H., and Winberg, J. (1997). Natural odor preferences of newborns change over time. Acta Paediatrica 86: 985–990.

Von Uexküll, J. (1957). A stroll through the worlds of animals and men. In: Instinctive Behaviour: The Development of a Modern Concept (eds. C. H. Schiller and K. S. Lashley), pp. 5–82. International Universities Press, Madison, Conn. (Original work published in 1934.)

Vygotsky, L. (1978). Mind in Society. Harvard University Press, Cambridge, Mass.

Wallace, B. (2010). Getting Darwin Wrong: Why Evolutionary Psychology Won't Work. Imprint Academic, Exeter.

Wanless, F. R. (1978). A revision of the spider genus, Portia (Araneae, Salticidae). Bulletin of the British Museum (Natural History) Zoology 34: 83–124.

Webb, B. (1995). Using robots to model animals: a cricket test. Robotics and Autonomous Systems 16: 117–134.

Webb, B. (1996). A cricket robot. Scientific American 275: 94–99.

Wellman, H. M., Cross, D., and Watson, J. (2001). Meta-analysis of theory of mind development: the truth about false belief. Child Development 72: 655–684.

Wells, A. (2002). Gibson's affordances and Turing's theory of computation. Ecological Psychology 14: 140–180.

Wells, A. (2006). Rethinking Cognitive Computation: Turing and the Science of the Mind. Palgrave, London.

Wheeler, M. (2005). Reconstructing the Cognitive World: The Next Step. MIT Press, Cambridge. Mass.

White, B. W., Saunders, F. A., Scadden, L., Bach-y-Rita, P., and Collins, C. C. (1970). Seeing with the skin. Perception and Psychophysics 7: 23–27.

White, P. A. (1995). The Understanding of Causation and the Production of Action. Erlbaum, Mahwah, N.J.

Whiting, B., and Barton, R. A. (2003), The evolution of the cortico-cerebellar complex in primates: anatomical connections predict patterns of correlated evolution. Journal of Human Evolution 44: 3–10.

Wicker, B., Keysers, C., Plailly, J., Royet, J. P., Gallese, V., and Rizzolatti, G. (2003). Both of us disgusted in my insula: the common neural basis of seeing and feeling disgust. Neuron 40: 655–664.

Wilcox, R. S., Jackson, R. R., and Gentile, K. (1996). Spiderweb smokescreen: spider trickster uses background noise to mask stalking movements. Animal Behaviour 51: 313–326.

Wilson, R. A., and Clark, A. (2009). How to situate cognition: letting nature takes its course. In: The Cambridge Handbook of Situated Cognition (eds. P. Robbins and M. Aydedem), pp. 55–78. Cambridge University Press, Cambridge.

Wilson, T. (2004). Strangers to Ourselves: Discovering the Adaptive Unconscious. Harvard University Press, Belknap Press, Cambridge, Mass.

Wozniak, R. H. (1993). Conwy Lloyd Morgan, mental evolution, and the Introduction to comparative psychology (introduction to republication). In: Morgan, C. L., Introduction to comparative psychology, pp. vii–xix. Routledge/Thoemmes Press, London. (Original work published 1894.)

Xu, Y. D., Liu, J., and Kanwisher, N. (2005). The M170 is selective for faces, not for expertise. Neuropsychologia 43: 588–597.

Yokoi, H., Arieta, A. H., Katoh, R., Yu, W., Watanabe, I., and Maruishi, M. (2004). Mutual adaptation in a prosthetics application. In: Embodied Artificial Intelligence, LNAI 3139 (eds. F. Iida, R. Pfeifer, L. Steels, and Y. Kuniyoshi), pp. 146–159. Springer-Verlag, Berlin.

Yurk, H., Barrett-Lennard, L., Ford, J.K.B., and Matkin, C. O. (2002). Cultural transmission within maternal lineages: vocal clans in resident killer whales in southern Alaska. Animal Behaviour 63: 1103–1119.

Zelazo, P. R. (1983). The development of walking: new findings and old assumptions. Journal of Motor Behavior 15: 99–137.

Zelazo, P. R., Zelazo, N.A., and Kolb, S. (1972). "Walking" in the newborn. Science 177: 1058–1059.

INDEX